T0298696

Operations Research

Advanced Research in Reliability and System Assurance Engineering

Series Editor: Mangey Ram

Professor, Graphic Era University, Uttarakhand, India

Reliability Engineering
Methods and Applications
Edited by Mangey Ram

Reliability Management and Engineering
Challenges and Future Trends
Edited by Harish Garg and Mangey Ram

Applied Systems Analysis
Science and Art of Solving Real-Life Problems
F. P. Tarasenko

Stochastic Models in Reliability Engineering
Lirong Cui, Ilia Frenkel, and Anatoly Lisnianski

Predictive Analytics
Modeling and Optimization
Vijay Kumar and Mangey Ram

Design of Mechanical Systems Based on Statistics
A Guide to Improving Product Reliability
Seong-woo Woo

Social Networks
Modeling and Analysis
Niyati Aggrawal and Adarsh Anand

Operations Research
Methods, Techniques, and Advancements
Edited by Amit Kumar and Mangey Ram

For more information about this series, please visit: https://www.routledge.com/
Advanced-Research-in-Reliability-and-System-Assurance-Engineering/book-
series/CRCARRSAE

Operations Research

Methods, Techniques, and Advancements

Edited by

Amit Kumar and Mangey Ram

CRC Press is an imprint of the
Taylor & Francis Group, an **informa** business

First edition published 2022
by CRC Press
6000 Broken Sound Parkway NW, Suite 300, Boca Raton, FL 33487-2742

and by CRC Press
4 Park Square, Milton Park, Abingdon, Oxon, OX14 4RN

Library of Congress Cataloging-in-Publication Data

Names: Kumar, Amit, editor. | Ram, Mangey, editor.
Title: Operations research: methods, techniques, and advancements/edited by Amit
    Kumar and Mangey Ram.
Other titles: Operations research (CRC Press)
Description: First edition. | Boca Raton, FL: CRC Press, 2022. | Series: Advanced
    research in reliability and system assurance engineering | Includes bibliographical
    references and index. | Summary: "This era of science and engineering has attracted
    researchers tasked with evaluating performance and optimization of problems in
    the field of operations research. The book covers mathematical analysis, methods
    and applications involving processes such as system performance, optimization,
    inventory theory, reliability theory, and queueing theory. Operations Research:
    Methods, Techniques, and Advancements explores recent and innovative methods
    and advancements associated with the mathematical theory of operations research.
    It offers a detailed overview of mathematical modelling for general industrial
    systems and emphasizes the latest ideas for the benefit of society and the research
    community. Intended for a broad range of readers, this book is useful to academicians,
    industrialists, researchers, students, academia and specialists from various disciplines
    and those working in the industry"— Provided by publisher.
Identifiers: LCCN 2021050491 (print) | LCCN 2021050492 (ebook) |
    ISBN 9780367741556 (hbk) | ISBN 9780367741570 (pbk) | ISBN 9781003156291 (ebk)
Classification: LCC T57.6 .O6456 2022 (print) | LCC T57.6 (ebook) |
    DDC 658.4/034—dc23/eng/20220106
LC record available at https://lccn.loc.gov/2021050491
LC ebook record available at https://lccn.loc.gov/2021050492

ISBN: 978-0-367-74155-6 (hbk)
ISBN: 978-0-367-74157-0 (pbk)
ISBN: 978-1-003-15629-1 (ebk)

DOI: 10.1201/9781003156291

Typeset in Times
by KnowledgeWorks Global Ltd.

Contents

v

Preface

For many decades, operations research (OR) has been used widely for decision-making in many of the relevant fields. It is one of the branches of mathematics which is generally implicated in many real-life industrial as well as theoretical problems through a variety of methods and techniques, e.g., mathematical modelling, simulation, numerical methods, game theoretic approach, system's reliability approach, queuing model, inventory model, and many more; hence, OR methods and their applications cannot be ignored when dealing with almost any kind of industrial/management problem. Keeping this in mind, this book contains various methods and techniques associated with various problems in the field of OR from a research standpoint, which are very much needed and helpful for scholars, practitioners, researchers, and academicians who have a keen interest in the associated fields.

This book has a vital role for an ample range of readers, including academicians, industrialist, researchers, and, last but not least, students from various disciplines, to learn about new concepts and research. A clear description is provided for each and every result, with the aid of graphs (whenever necessary). It is, therefore, not only useful for academia but also provides a rich orientation for specialists in the relevant industry.

The book consists of ten chapters. Chapter 1 presents a novel approach for universal generating function (UGF) implementation through Excel software for performance evaluation of a multi-state system. In Chapter 2, the author considers the traditional age replacement policy for some models, such as the bridge structure and the r-within-consecutive-k-out-of-n: F system, with independent and identically distributed components under some specific conditions. Chapter 3 presents the discussion with the quickest path reliability problem and then studies the disjoint minimal paths' reliability problems with two and k disjoint minimal paths. For each problem, an algorithm is identified and illustrated by using a benchmark network example. In Chapter 4, optimization of a nonlinear innovation diffusion model with three different compartments is proposed. A TODIM-based multi-criteria decision-making approach is proposed for evaluating the optimal inventory policy under the Pythagorean Fuzzy Environment in Chapter 5. Moreover, this chapter is also illustrating one such algorithm to determine the best inventory control policy to be practiced based on some major criteria. Chapter 6 demonstrates how one can obtain the different reliability measures for a system having a k-out-of-n redundancy by using mathematical modelling and Markov birth death process. Chapter 7 presents a study in which the authors minimize the discounted cash flow of weighted delay penalty costs in resource-constrained project scheduling problems through three different metaheuristic approaches, including imperialist competitive algorithm, biogeography-based optimization algorithm, and the shuffled frog leaping algorithm. Chapter 8 investigates a profit maximization inventory control model with partial backlogging and deterioration in demand. Also, in this problem, the author has considered the rate of production as a decision variable where the cost of one-unit production is assumed as a function of the production rate. Chapter 9 introduces the Joint Birnbaum-Reliability

Achievement Worth, Joint Birnbaum-Reliability Reduction Worth, Joint Birnbaum-Reliability Fussell-Vesely measure, and Analogues Joint Risk importance measures for three multistate components of a multistate system. Last but not least, the importance of reliability for enhancing the performance of a system is discussed in Chapter 10. In this chapter, the authors also discuss the effect of reliability on performance, measuring methods of equipment reliability, and available approaches for increasing the reliability of a system.

Editors
Amit Kumar
Pune, Maharashtra, India

Mangey Ram
Uttarakhand, India

Acknowledgments

The editors acknowledge CRC Press (Taylor & Francis Group) for this opportunity and professional support. Also, we would like to thank all the chapter authors and reviewers for their availability for this project.

Editors

Dr. Amit Kumar is an assistant professor in the Department of Mathematics at Symbiosis Institute of Technology, Symbiosis International University, Pune, India. He has taught several core courses in pure and applied mathematics at undergraduate and postgraduate levels. He has received his bachelor's and master's from Chaudhary Charan Singh University Meerut, India, in 2006 and 2009, respectively. In 2016, he completed his doctorate in applied mathematics from Graphic Era (Deemed to be University), Dehradun Uttarakhand, India, in the field of reliability theory. He published several research papers/books/book chapters in various esteemed international journals/books including Taylor & Francis, Springer, Emerald, World Scientific, InderScience, and many other national and international journals of repute, and also presented his works at national and international conferences. He is a reviewer of many international journals including Elsevier, Springer, Emerald, John Wiley, Taylor & Francis, and more. His fields of research are operations research, reliability theory, fuzzy reliability, and systems engineering.

Dr. Mangey Ram received a PhD in mathematics and computer science from G.B. Pant University of Agriculture and Technology, Pantnagar, India. He has been a faculty member for twelve years and has taught several core courses in pure and applied mathematics at undergraduate, postgraduate, and doctorate levels. He is currently a research professor at Graphic Era (Deemed to be University), Dehradun, India. Before joining Graphic Era, he was a deputy manager (probationary officer) with Syndicate Bank for a short period. He is editor-in-chief of *International Journal of Mathematical, Engineering and Management Sciences*, and guest editor and editorial board member of various journals. He is a regular reviewer for international journals including IEEE, Elsevier, Springer, Emerald, John Wiley, Taylor & Francis, and many others. He has published more than 200 research publications in IEEE, Taylor & Francis, Springer, Elsevier, Emerald, World Scientific, and many other national and international journals of repute, and has also presented his works at national and international conferences. His fields of research are reliability theory and applied mathematics. Dr. Ram is a senior member of the IEEE, life member of Operational Research Society of India, Society for Reliability Engineering, Quality and Operations Management in India, Indian Society of Industrial and Applied Mathematics, member of International Association of Engineers in Hong Kong, and Emerald Literati Network in the United Kingdom. He has been a member of the organizing committee of a number of international and national conferences, seminars, and workshops. He has been conferred with "Young Scientist Award" by the Uttarakhand State Council for Science and Technology, Dehradun, in 2009. He has been awarded the "Best Faculty Award" in 2011, "Research Excellence Award" in 2015, and "Outstanding Researcher Award" in 2018 for his significant contribution in academics and research at Graphic Era (Deemed to be University), Dehradun, India.

Contributors

Fatemeh Afsharnia
Department of Agricultural Machinery
  and Mechanization Engineering
Agricultural Sciences and Natural
  Resources University of Khuzestan
Ahvaz, Iran

Sayyid Ali Banihashemi
Department of Industrial Engineering
Payame Noor University
Tehran, Iran

V.M. Chacko
Department of Statistics
St. Thomas College (Autonomous)
Thrissur, University of Calicut
Kerala, India

Majid Forghani-elahabad
Center of Mathematics, Computing,
  and Cognition
Federal University of ABC
Santo André, SP, Brazil

Prabjot Kaur
Birla Institute of Technology Mesra
Ranchi, India

Mohammad Khalilzadeh
CENTRUM Católica Graduate
  Business School, Lima, Peru
Pontificia Universidad Católica del Perú
Lima, Peru

Amit Kumar
Department of Mathematics
Symbiosis Institute of Technology
Symbiosis International University
Pune, India

Pardeep Kumar
Department of Mathematics
Faculty of Science
Lovely Professional University
Phagwara, Punjab, India

Rakesh Kumar
Department of Applied Sciences
Shaheed Bhagat Singh State
  University
Ferozepur, Punjab, India

Anjali Priya
Department of Mathematics
Birla Institute of Technology
  Mesra
Ranchi, India

Mangey Ram
Department of Mathematics
Computer Science & Engineering
Graphic Era Deemed to be
  University
Dehradun, Uttarakhand

Sanjay Sharma
Department of Applied Sciences
  & Humanities
Ajay Kumar Garg Engineering
  College
Ghaziabad, India

Ioannis S. Triantafyllou
Department of Computer
  Science and Biomedical
  Informatics
University of Thessaly
Lamia, Greece

Anand Tyagi
Department of Mathematics
BKBIET
Pilani, India

B.B. Verma
Department of Applied Sciences
    & Humanities
Ajay Kumar Garg Engineering College
Ghaziabad, India

1 A Novel Approach for UGF Implementation through Excel Software for Behavior Analysis of a Multi-State System

Amit Kumar
Department of Mathematics, Symbiosis Institute of
Technology, Symbiosis International University, Pune, India

Pardeep Kumar
Department of Mathematics, Faculty of Science,
Lovely Professional University, Phagwara, Punjab, India

CONTENTS

1.1 INTRODUCTION

With the advent of computer technology, these days many good softwares are available in the market like MATLAB, MAPLE, etc., which can perform the computation very quickly and precisely. These softwares are capable of performing very complex calculations in very little time. But for using these softwares, one must have to have

DOI: 10.1201/9781003156291-1

sound knowledge of the commands of these softwares and the computer programming. It is not everyone's cup of tea to learn computer programming. Hence, it is a great hurdle for new researchers or for those people who are not computer savvy. These people spend a great amount of time in learning these softwares to be able to get research-related results. Therefore, this is the need of the hour to have an easy way of implementing the technique on the other software, which gives the same results with great ease. It helps the researcher to spend more time learning other research-related activities. Here, in this chapter, authors suggest a new approach to implement the UGF technique on Excel software which gives the same result which one can obtain with the help of algebraic multiplication of the UGF polynomials of the system's component.

In every industry, many machines are installed to produce products. Some industrial systems are very complex and have thousands of components. These components work with varying degree of performance levels and have multiple failure modes, which affect the system performance level. The system which works with a varying level of performance is called a multistate system (MSS). In an MSS, the system and its components both have different performance levels. Many engineering systems like power systems, flow transmission systems and data transmission systems work with varying degrees of performance levels. For example, in a power system, a generator may be capable of producing electricity equal to 150 MW, but due to the degradation in the generator components, it may be capable of producing electricity equal to 100 MW. Further, degradation in the generator components may cause the generator to produce electricity equal to 50 MW. In the end, when components degrade further, it fails completely. So, here this generator works with four performance levels $\{150\,\text{MW}, 100\,\text{MW}, 50\,\text{MW}, 0\,\text{MW}\}$. So, in this way, if there are four generators installed in the power station, it is necessary to know that at time "t" system is working with what performance rate? So, for this, four methods have been purposed in the literature for the reliability indices determination: (1) structure-function approach, (2) stochastic-process approach, (3) universal generating function (UGF) technique and (4) Monte Carlo simulation technique. The structure-function method technique can be used when the state of each component is known. But for a system with a large number of components, this technique is not preferred. Stochastic-process approach suffers from dimensionality curse. Even for a small system, the number of system states increases very drastically and one may easily forget to take any of the system states in the transition state diagram. This is the main disadvantage of the stochastic-process method. Monte Carlo simulation technique is based on the simulations, and therefore, sometimes it also takes a long time to get the simulation results. The last technique is the UGF technique, which has been extensively used by the researchers for obtaining various performance measures of the MSS. But the implementation of this technique can be improved. In the next section, we give the literature of the UGF technique and MSS.

1.2 ASSOCIATED LITERATURE

As it is quite clear, UGF technique is used for the reliability evaluation of the MSS.

First of all, this technique was developed by Igor Ushakov [1]. She used this technique in the redundancy optimization problem. Later, this technique was used in the reliability evaluation of the MSS. She used the Descartes product method to determine the reliability indices. Later, Levitin [2] contributed a lot to the reliability evaluation of the MSS's performance analysis. He developed various methods and algorithms for the evaluation of the reliability performance indicator of the MSS. Levitin and Lisnianski [3] presented the sensitivity and important analysis of the MSS, which helps in determining the most critical components of the system. Lisnianski et al. [4] also applied the UGF technique in the reliability evaluation of the power system. Wen et al. [5] used the UGF technique in the reliability evaluation of the compressor system. This has been observed that some systems' components on the failure, also, destroy the other components of the system. This analysis was presented by Levitin and Xing [6]. They showed in their research paper that the failure propagation in the system may destroy the other components of the system. In the development of the theory of the MSS, the algorithms for determining the reliability of k-out-of-n were developed. Chaturvedi et al. [7] developed an efficient algorithm for the reliability determination of the k-out-of-n system. This algorithm gives the exact reliability of the large MSS. Yingkui and Jing [8] presented the systematic literature review of the all the literature available of the MSS up to the year 2012. The major problem of the UGF technique is that it determines the performance distribution of the system from the known performance distribution of its components. Lisnianski [9] studied the dynamic MSS with the help of L_z-transformation. He presented the various properties of the L_z-transformation and discussed a few examples. In real-life situation, sometimes the precise and accurate data collection is a great hurdle. Due to the insufficient data, precise probabilities of the states cannot be determined. In this case, Dempster–Shafer's theory is used to determine the reliability indices of an MSS. The use of the theory is presented by Refs. [10], [11] in their research papers. Meenakshi and Singh [12] further extended the work and applied it on the $((e, f), k, I_C) / (m, n)) : F$ system where the system can fail if (e, f) sub-matrix fails or any k components fail or consecutive I_c components fail in m rows or in n columns of a (m, n) sub-matrix. Probabilities of the components were obtained with the help of stochastic Markov modeling. After this, reliability, mean time to failure (MTTF) was obtained for the considered system. Also, sensitivity analysis was performed to determine the most critical components of the system. Meena and Vasanthi [13] discussed the use of the MANET (mobile ad hoc network) in the battlefield because it is easy to use wireless communication systems in the battlefield rather than using the fixed wire network on the battlefield. He used the new modified UGF technique to determine the reliability of the MANETs system. Nair and Manoharan [14] used L_z-transformation for the evaluation of the power system, which was given minimal repair when the system's components fail. In an MSS, if two systems are considered, the performance of the systems can be shared with each other. In these systems, when one system satisfies its own demand and has surplus performance, then that surplus performance can be shared with the other system. Levitin [15] and Wang et al. [16] in their work showed how the performance of the system can be shared. For this, they used the UGF technique. Dong et al. [17] presented in their research that when the connection between system components is uncertain, then the weighted UGF technique can be used for the reliability evaluation of the system. In their research, they compared the various performance measures of the MSS for flow transmission system, task processing system, fixed weighted MSS and variable weighted MSS. For getting more detail on MSS, one can refer to Refs. [18]–[20]. Ding and Lisnianski [21] presented the concept of fuzzy universal technique for the reliability evaluation of the MSS.

It is quite clear from the above discussion that the UGF technique is the best technique for the reliability evaluation of the MSS. But its implementation without the proper knowledge of computer programming is very difficult. All the researchers in the above literature used MATLAB software for the performance evaluation of the MSS using the UGF technique or used the Descartes product rule. To the best of the author's knowledge, no one has ever tried to implement the UGF technique on the Excel software, which is very easy and can save a lot of time for the research community. Therefore, the authors in this chapter present how the UGF technique can be implemented on Excel software very easily. The remainder of the chapter is organized as follows: in Section 1.3, various notations used in the paper are given. In Section 1.4, the UGF definition and its various composition operators are given. In Section 1.5, multistate performance indices are described. In Section 1.6, implementation of the UGF technique on Excel software is given. In Section 1.7, result and discussion are given.

1.3 NOTATIONS

The following notations (given in Table 1.1) will be used in the chapter for understanding the mathematical calculations.

TABLE 1.1
Notations

N	Number of system components
$k_i\ (i = 1,2...N)$	Number of states of the ith component
$G_i(t)\ (i = 1,2...,N)$	Represents performance rate variable of the ith component at time t
$g_i\ (i = 1,2...N)$	Random variable representing the performance of the ith component
g_{ij}	Represents the performance rate of the ith component in the jth state
p_{ij}	Represents the probability of the ith component in the jth state
$u(z)$	Represents the UGF of the system
$u_i(z)$	Represents the UGF of the ith component of the system
A	Represents system availability
δ_A	Represents availability operator
E	Represents mean output performance
D	Represents mean output performance deficiency
I	Represents indicator function
k	Total number of system states
f	Structure-function
MSS	Multistate system

1.4 UGF AND VARIOUS COMPOSITE OPERATORS

Suppose that a system has "N" elements, where the ith element of the system has different performance rates, which is represented by the performance rate variable $G_i(t)$ that takes the value from the set $g_i = \{g_{i1}, g_{i2}, g_{i3} \ldots g_{ik_i}\}$, where g_{ij} is the performance rate of the ith component in the jth state. The probabilities associated with ith element in the various states are given by the set $p_i = \{p_{i1}, p_{i2}, p_{i3} \ldots p_{ik_i}\}$, where p_{ij} is the probability of the ith component in the jth state. Therefore, UGF polynomial for the ith element is given by:

$$u_i(z) = \sum_{j=1}^{k_i} p_{ij}\, z^{g_{ij}} \tag{1.1}$$

In this way, the UGF polynomial of every element is written. In order to obtain various performance indicators of the MSS, the UGF polynomial of the whole system is obtained which is given below.

$$u(z) = \sum_{j_1=1}^{k_1} \sum_{j_2=1}^{k_2} \cdots \sum_{j_N=1}^{k_N} p_{ij_1}\, p_{ij_2} \ldots p_{ij_N}\, z^{f(g_{ij_1}, g_{ij_2} \ldots g_{ij_N})} \tag{1.2}$$

This f is called the structure-function. There are basically two types of structure-function: f_{series} and $f_{parallel}$. f_{series} and $f_{parallel}$ give different performance rate values, depending upon which type of system we are dealing with.

For a flow transmission system:

$$f_{series} = \min(G_1(t), G_2(t) \cdots G_n(t)) \tag{1.3}$$

For a task processing system:

$$f_{series} = \cfrac{1}{\cfrac{1}{G_1(t)} + \cfrac{1}{G_2(t)} + \cdots + \cfrac{1}{G_n(t)}} \tag{1.4}$$

For a flow transmission system:

$$f_{parallel} = (G_1(t) + G_2(t). + \cdots + G_n(t)) \tag{1.5}$$

For a task processing system:

$$f_{parallel} = \max(G_1(t), G_2(t) \cdots G_n(t)) \tag{1.6}$$

From Equation (1.2), it is quite obvious that a total number of system states are equal to:

$$k = \prod_{i=1}^{N} k_i \tag{1.7}$$

But the fact is that some system states have the same performance rate values, then we combine the terms which have the same performance rate and it just becomes only one state of the system. In this way, the UGF technique reduces the total number of system states. To make things more clear, we consider a flow transmission system with two elements. Suppose for the first element, the performance rate set is $g_1 = \{g_{11}, g_{12}, g_{13}\}$ and the corresponding probability set is $p_1 = \{p_{11}, p_{12}, p_{13}\}$ and for the second element, the performance rate set is $g_2 = \{g_{21}, g_{22}\}$ and the corresponding probability set is $p_2 = \{p_{21}, p_{22}\}$. Therefore, the UGFs of first and second elements are given by:

$$u_1(z) = p_{11} z^{g_{11}} + p_{12} z^{g_{12}} + p_{13} z^{g_{13}} \tag{1.8}$$

$$u_2(z) = p_{21} z^{g_{21}} + p_{22} z^{g_{22}} \tag{1.9}$$

When these components work in series, then the UGF of the whole system can be obtained as:

$$u(z) = u_1(z) \underset{series}{\otimes} u_2(z)$$

$$u(z) = \left(p_{11} z^{g_{11}} + p_{12} z^{g_{12}} + p_{13} z^{g_{13}} \right) \underset{series}{\otimes} \left(p_{21} z^{g_{21}} + p_{22} z^{g_{22}} \right)$$

$$\begin{aligned} u(z) = \; & p_{11} \cdot p_{21} z^{\min(g_{11}, g_{21})} + p_{11} \cdot p_{22} z^{\min(g_{11}, g_{22})} + p_{12} \cdot p_{21} z^{\min(g_{12}, g_{21})} \\ & + p_{12} \cdot p_{22} z^{\min(g_{12}, g_{22})} + p_{13} \cdot p_{21} z^{\min(g_{13}, g_{21})} + p_{13} \cdot p_{22} z^{\min(g_{13}, g_{22})} \end{aligned} \tag{1.10}$$

When these components work in parallel, then the UGF of the whole system can be obtained as:

$$u(z) = u_1(z) \underset{parallel}{\otimes} u_2(z)$$

$$u(z) = \left(p_{11} z^{g_{11}} + p_{12} z^{g_{12}} + p_{13} z^{g_{13}} \right) \underset{parallel}{\otimes} \left(p_{21} z^{g_{21}} + p_{22} z^{g_{22}} \right)$$

$$\begin{aligned} u(z) = \; & p_{11} \cdot p_{21} z^{(g_{11}+g_{21})} + p_{11} \cdot p_{22} z^{(g_{11}+g_{22})} + p_{12} \cdot p_{21} z^{(g_{12}+g_{21})} \\ & + p_{12} \cdot p_{22} z^{(g_{12}+g_{22})} + p_{13} \cdot p_{21} z^{(g_{13}+g_{21})} + p_{13} \cdot p_{22} z^{(g_{13}+g_{22})} \end{aligned} \tag{1.11}$$

From Equations (1.10) and (1.11), it is quite clear that for a small system with two components where component 1 has three performance rates and component 2 has only two performance rates, the whole system has six performance states or less depending on the values of the performance rates values of the component 1 and component 2. If in the system number of components are more, these polynomial terms increase very rapidly and become unmanageable when it is implemented on the MATLAB or solved manually this is the main drawback of this technique. But its implementation on Excel software is very easy. New research scholars and those who are not computer savvy can easily implement it on the Excel software with ease and can easily obtain the desired results. Here, also notice one more thing in Equation (1.10), when the exponent of z is $\min(g_{11}, g_{21})$, then system state probability is $p_{11} \cdot p_{21}$. Similarly in Equation (1.11), when the exponent of z is $(g_{11} + g_{21})$, then system state probability is $p_{11} \cdot p_{21}$. Hence, we establish the relationship between exponents of z and its corresponding probabilities. In general, if the exponent of z is $\min(g_{11}, g_{21}, ..., g_{N1})$, then its corresponding probability is $p_{11} \cdot p_{21} ... p_{N1}$. Similarly, if the exponent of z is

$(g_{11} + g_{21} + \cdots + g_{N1})$, then its corresponding probability is $p_{11}.p_{21}...p_{N1}$. For the computations of this chapter, this argument will be used. In the next section, we give a description of the various performance measures of the MSS.

1.5 MULTISTATE SYSTEM PERFORMANCE INDICATORS

Once the final UGF of the whole system is obtained, then the system may have many performance rates. Only those states are considered good states of the system whose performance rate is greater than or equal to a certain demand w. Therefore, the whole system is divided into two subsets: acceptable and non-acceptable. The states which satisfy $r_i = g_i - w \geq 0$ come in the acceptable subset and the states which satisfy $r_i = g_i - w \leq 0$ come in the unacceptable subset.

1.5.1 AVAILABILITY OF THE SYSTEM

Availability of the MSS is defined as the probability that MSS performance rate is greater than w. Mathematically, it can be expressed as:

$$A = \sum_{r_i \geq 0} p_i \tag{1.12}$$

Using availability operator, availability of the system can be represented as:

$$A = \delta_A(U(z), w) = \sum_{i=1}^{n} p_i.\beta_i \tag{1.13}$$

where

$$\beta_i = \begin{cases} 1 & r_i \geq 0 \\ 0 & r_i \prec 0 \end{cases}$$

1.5.2 MEAN OUTPUT PERFORMANCE

Mean output performance of the system is defined as the average value of the output performance of the system. Mathematically, it can be obtained as:

$$E = \sum_{i=1}^{k} g_i.p_i \tag{1.14}$$

1.5.3 MEAN OUTPUT PERFORMANCE DEFICIENCY

The mean output performance deficiency of the MSS is average value when the system output performance doesn't satisfy its demand. Mathematically, it can be expressed as:

$$D = \sum_{r_i \leq 0} p_i(w - g_i) \tag{1.15}$$

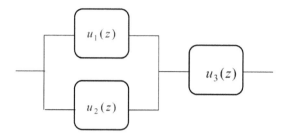

FIGURE 1.1   A flow transmission system structure.

1.6   IMPLEMENTATION OF UGF TECHNIQUE ON THE EXCEL SOFTWARE

In order to show the implementation of the UGF technique on Excel software, we consider here one flow transmission system having three multistate elements as shown in the following diagram.

The rate of this flow transmission is taken in tons/per minute. The first element has the performance rates given by the set $g_1 = \{g_{11}, g_{12}, g_{13}\}$ and the corresponding probabilities are given by the set $p_1 = \{p_{11}, p_{12}, p_{13}\}$. The performance rate of the second element is given by the set $g_2 = \{g_{21}, g_{22}, g_{23}\}$ and the corresponding probabilities are given by the set $p_2 = \{p_{21}, p_{22}, p_{23}\}$. Similarly, the performance rate of the third element is given by the set $g_3 = \{g_{31}, g_{32}\}$ and the corresponding probabilities are given by the set $p_3 = \{p_{31}, p_{32}\}$. The diagram of the system is presented in Figure 1.1.

Here, element 1 and element 2 work in the parallel configuration and element 3 works in a series configuration with the first two elements. Enter the performance rate of the first element in column A, performance rates of the second element in column B and performance rate of the third element in column C. Select the first column A alone and from the keyboard press CTRL+T keys combination then "Create Table box" appears, then check the box "My table has headers" and click OK button. Do the same step for the column B. For the column C, the step is shown in the following screenshot.

Step 1:

C1		fx	Element 3 Performance rates			
	A		B		C	D
1	Element 1 Performance rates		Element 2 Performance rates		Element 3 Performance rates	
2	g11		g21		g31	
3	g12		g22		g32	
4	g13		g23			

Create Table   ?   ×
Where is the data for your table?
=SCS1:SCS4
☑ My table has headers
OK   Cancel

Once it has been done for the all three columns, we get the following screen.

Step 2:

After this step, select any cell of column A except the heading of column A and from the keyboard press ALT+N+V keys combination. The "Create Pivot Table" dialogue box appears, click on the check box "Existing worksheet" and select any blank location of the worksheet. Here authors selected the D2 cell, then this location appears in the "Location box". After this step, tick on the check box "Add this data to the Data Model". The following screenshot explains it completely, then click the OK button.

Step 3:

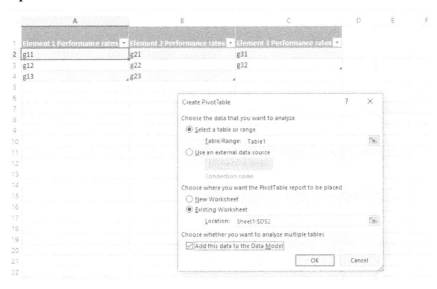

After this, the following screen appears.

Step 4:

In the above screen on upper right corner of the screen pivot table field is written. Below it, there are two categories: "Active" and "All". By default, the "Active" field is active. Below it, Table 1 is given. Click the check box of "Element 1 performance rate". In this way, Table 1 data is fitted in the model. After this click on the "All" field, the following screen appears.

Step 5:

On the right-hand side of the screen, Table 2 and Table 3 are written. Click on Table 2 and then tick the check box of element 2 performance rate, do the same for Table 3. In this way, Table 2 data and Table 3 data are also added to the model. The following screen appears after this step.

Step 6:

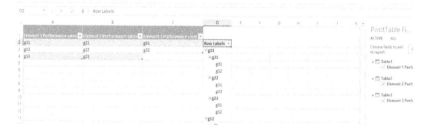

After this, click on the Design button from the menu bar. It is shown in the following screenshot.

Step 7:

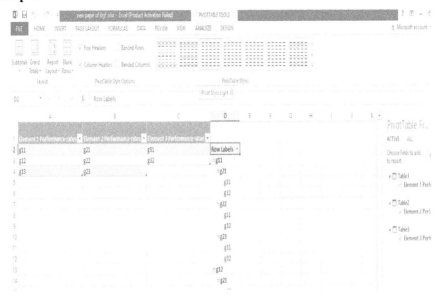

Now click on the "Report Layout", a drop-down menu appears as shown in the following screenshot.

Step 8:

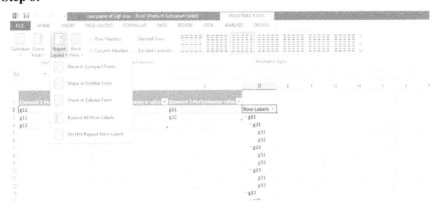

Now click on the show tabular form, the screen appears like this as shown in the following screenshot.

Step 9:

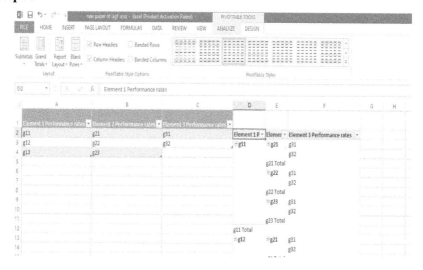

After this step, again click on the "Report Layout" and click on the "Repeat All Item Labels", as it is shown in the following screenshot.

Step 10:

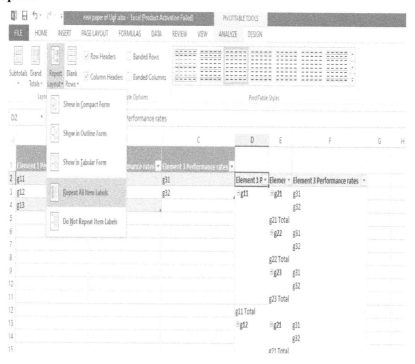

After this step, the following screen appears. The screenshot of which is given below.

Step 11:

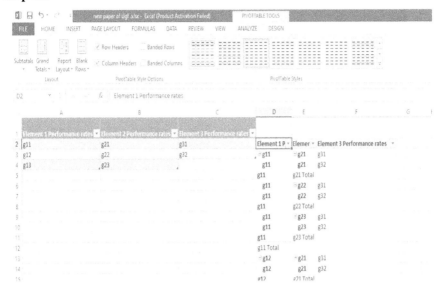

After this step, click on the "Grand Totals", as it is shown in the following screenshot.

Step 12:

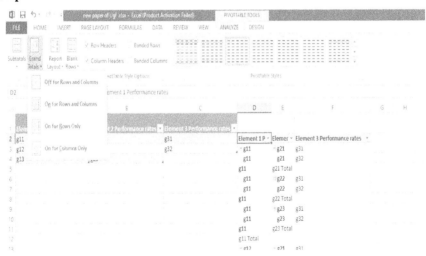

Click on the option "off for Rows and Columns", the following screen appears.

Step 13:

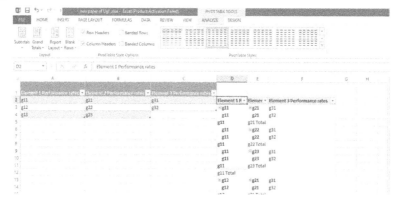

After this step, click on the subtotals as shown in the following screenshot.

Step 14:

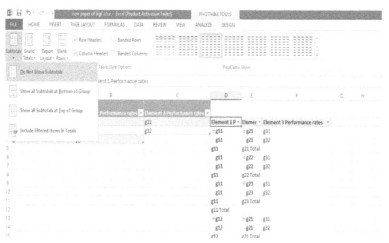

Click on the option "Do Not Show Subtotals". The following screen appears.

Step 15:

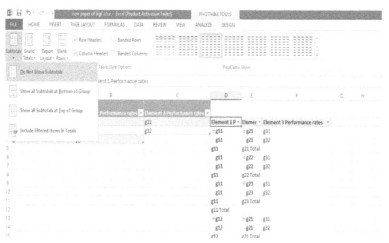

Columns D, E and F represent the total possible combination of the performance rates of the system. Once we have a total combination of the performance rates of the system, we can apply operations as per the configuration of the system. Also one can easily write the corresponding state probabilities. For example, when the combination of the performance rate of system elements is (g_{11}, g_{21}, g_{31}), then the corresponding probability is $P_{11}.P_{21}.P_{31}$. Similarly, when the combination of the performance rate of the system elements is (g_{11}, g_{21}, g_{32}), then the corresponding probability is $P_{11}.P_{21}.P_{32}$. This is shown in the following screenshot.

Step 16:

As these are not numerical values so it is not possible to show the computation procedure. Next, we show you how computation is performed.

Consider the first element has the performance rates $g_{11} = 1.5$, $g_{12} = 1$, $g_{13} = 0$ and the corresponding probabilities are $P_{11} = 0.8$, $P_{12} = 0.1$, $P_{13} = 0.1$. The second element $g_{21} = 2$, $g_{22} = 1.5$, $g_{23} = 0$ and the corresponding probabilities are $P_{21} = 0.7$, $P_{22} = 0.2$, $P_{23} = 0.1$. The third element has the performance rates $g_{31} = 4$, $g_{32} = 0$ and the corresponding probabilities are $P_{31} = 0.96$, $P_{32} = 0.04$. Enter this data in the same excel sheet we get the following screen.

Step 17:

	A	B	C	D	E	F	G
1	Element 1 Performan	Element 2 Performance	Element 3 Performar	G1	G2	G3	
2	1.5	2		4 Element 1 P	Elemer	Eleme	Probability
3	1	1.5	0	1.5	2	4	0.5376
4	0	0		1.5	2	0	0.0224
5				1.5	1.5	4	0.1536
6				1.5	1.5	0	0.0064
7				1.5	0	4	0.0768
8				1.5	0	0	0.0032
9				1	2	4	0.0672
10				1	2	0	0.0028
11				1	1.5	4	0.0192
12				1	1.5	0	0.0008
13				1	0	4	0.0096
14				1	0	0	0.0004
15				0	2	4	0.0672
16				0	2	0	0.0028
17				0	1.5	4	0.0192
18				0	1.5	0	0.0008
19				0	0	4	0.0096
20				0	0	0	0.0004

As computation is still not possible because these are not simple data values, now select the whole data and copy this data to "Sheet 2". After pasting this data on the second sheet again select the data, then on the right-hand side "Ctrl" appears, click on the down arrow and then select the paste special option as shown in the following screenshot.

Step 18:

			G1	G2	G3	Probability
Element 1 Performance rat	Element 2 Performa nce rat	Element 3 Performa nce rat	4 Elemer	Elemer	Elemer	
1.5	2		1.5	2	4	0.5376
1	1.5	0	1.5	2	0	0.0224
0	0		1.5	1.5	4	0.1536
			1.5	1.5	0	0.0064
			1.5	0	4	0.0768
			1.5	0	0	0.0032
			1	2	4	0.0672
			1	2	0	0.0028
			1	1.5	4	0.0192
			1	1.5	0	0.0008
			1	0	4	0.0096
			1	0	0	0.0004
			0	2	4	0.0672
			0	2	0	0.0028
			0	1.5	4	0.0192
			0	1.5	0	0.0008
			0	0	4	0.0096

Paste

Paste Values

Paste Options

Values (V)

Sheet1   Sheet2

After this, the following screen appears.

Step 19:

	A	B	C	D	E	F	G	H
1	Element 1	Element 2	Element 3	G1	G2	G3		
2	1.5	2	4	Element 1	Element 2	Element 3	Probability	
3	1	1.5	0	1.5	2	4	0.5376	
4	0	0		1.5	2	0	0.0224	
5				1.5	1.5	4	0.1536	
6				1.5	1.5	0	0.0064	
7				1.5	0	4	0.0768	
8				1.5	0	0	0.0032	
9				1	2	4	0.0672	
10				1	2	0	0.0028	
11				1	1.5	4	0.0192	
12				1	1.5	0	0.0008	
13				1	0	4	0.0096	
14				1	0	0	0.0004	
15				0	2	4	0.0672	
16				0	2	0	0.0028	
17				0	1.5	4	0.0192	
18				0	1.5	0	0.0008	
19				0	0	4	0.0096	
20				0	0	0	0.0004	
21								

Now select columns D, E and F. A yellow box appears in which an exclamation sign is written. Click on the drop-down arrow and then from these options click on the option "Convert to Numbers" as shown in the following screenshot.

Step 20:

	A	B	C	D	E	F	G	H	I
1	Element 1	Element 2	Element 3	G1	G2	G3			
2	1.5	2	4	Element 1	Element 2	Element 3	Probability		
3	1	1.5	! ▾	1.5	2	4	0.5376		
4				1.5	2	0	0.0224		
5	Number Stored as Text			1.5	1.5	4	0.1536		
6	Convert to Number			1.5	1.5	0	0.0064		
7	Help on this error			1.5	0	4	0.0768		
8	Ignore Error			1.5	0	0	0.0032		
9	Edit in Formula Bar			1	2	4	0.0672		
10				1	2	0	0.0028		
11	Error Checking Options...			1	1.5	4	0.0192		
12				1	1.5	0	0.0008		
13				1	0	4	0.0096		
14				1	0	0	0.0004		
15				0	2	4	0.0672		
16				0	2	0	0.0028		
17				0	1.5	4	0.0192		
18				0	1.5	0	0.0008		
19				0	0	4	0.0096		
20				0	0	0	0.0004		
21									

Now the whole Excel sheet contains data in which operations are permissible. The screen appears as given below.

Step 21:

Element 1	Element 2	Element 3	G1 Element 1	G2 Element 2	G3 Element 3	Probability
1.5	2	4	1.5	2	4	0.5376
1	1.5	0	1.5	2	0	0.0224
0	0		1.5	1.5	4	0.1536
			1.5	1.5	0	0.0064
			1.5	0	4	0.0768
			1.5	0	0	0.0032
			1	2	4	0.0672
			1	2	0	0.0028
			1	1.5	4	0.0192
			1	1.5	0	0.0008
			1	0	4	0.0096
			1	0	0	0.0004
			0	2	4	0.0672
			0	2	0	0.0028
			0	1.5	4	0.0192
			0	1.5	0	0.0008
			0	0	4	0.0096
			0	0	0	0.0004

Now we perform the operations for the configuration given in Figure 1.1. We get the following screenshot.

Step 22:

Element 1 Performance rates	Element 2 Performance rates	Element 3 Performance rates	G1 Element 1 Performance rates	G2 Element 2 Performance rates	G3 Element 3 Performance rates	H=G1+G2	K=MIN(G3,H)	Probability
1.5	2	4	1.5	2	4	3.5	3.5	0.5376
1	1.5	0	1.5	2	0	3.5	0	0.0224
0	0		1.5	1.5	4	3	3	0.1536
			1.5	1.5	0	3	0	0.0064
			1.5	0	4	1.5	1.5	0.0768
			1.5	0	0	1.5	0	0.0032
			1	2	4	3	3	0.0672
			1	2	0	3	0	0.0028
			1	1.5	4	2.5	2.5	0.0192
			1	1.5	0	2.5	0	0.0008
			1	0	4	1	1	0.0096
			1	0	0	1	0	0.0004
			0	2	4	2	2	0.0672
			0	2	0	2	0	0.0028
			0	1.5	4	1.5	1.5	0.0192
			0	1.5	0	1.5	0	0.0008
			0	0	4	0	0	0.0096
			0	0	0	0	0	0.0004

Column H gives the system performance rates of the system as one can see that the same value appears in many cells. To get the UGF of the system, find the sum of the probabilities of the same performance rates. For this purpose, one can also use the filter option. It is shown in the following screenshot for the performance level $K = 0$.

	A	B	C	D	E	F	G	H	I
	Element 1 Performance rates	Element 2 Performance rates	Element 3 Performance rates	G1	G2	G3		.T	
1									
4	0	0		1.5	2	0	3.5	0	0.0224
6				1.5	1.5	0	3	0	0.0064
8				1.5	0	0	1.5	0	0.0032
10				1	2	0	3	0	0.0028
12				1	1.5	0	2.5	0	0.0008
14				1	0	0	1	0	0.0004
16				0	2	0	2	0	0.0028
18				0	1.5	0	1.5	0	0.0008
19				0	0	4	0	0	0.0096
20				0	0	0	0	0	0.0004
21								sum	0.0496
22									
23									

Apply this filter can be used for obtaining the probability of the system state corresponding to its performance rate. Hence, the UGF of the whole system can be obtained as:

$$U(z) = 0.5376\,z^{3.5} + 0.2208\,z^{3} + 0.0192\,z^{2.5} + 0.0672\,z^{2} + 0.096\,z^{1.5} + 0.0096\,z^{1} + 0.0496\,z^{0}$$

Suppose that, system demand level is 1.5. Therefore, the availability of the system can be easily obtained from the system UGF.

$$A(1.5) = \delta_A(U(z), 1.5)$$

$$= 0.5376 + 0.2208 + 0.0192 + 0.0672 + 0.096$$

$$= 0.9408$$

Also, expected output of the performance can be calculated from the system UGF using the following formula:

$$E = \sum_{i=1}^{k} g_i \cdot p_i$$

$$= 0.5376 \times 3.5 + 0.2208 \times 3 + 0.0192 \times 2.5 + 0.0672 \times 2$$
$$+ 0.096 \times 1.5 + 0.0096 \times 1 + 0.0496 \times 0$$
$$= 2.88$$

System's mean output performance deficiency can be calculated from the system's UGF expression.

Using the following formula:

$$D = \sum_{\eta \leq 0} p_i(w - g_i)$$

$$= 0.5 \times 0.0096 + 1.5 \times 0.0496$$

$$= 0.0792$$

1.7   RESULT AND DISCUSSION

As for the performance evaluation of the MSS, the UGF technique is generally employed as it is just based on the simple algebraic multiplication. But its implementation on software like Maple, MATLAB is impossible without the proper knowledge of programming. Learning programming is not an easy task for everyone and it also takes time to learn programming. Also the Descartes product rule is very slow and not suitable method for large system. Hence, the authors of this chapter have discussed how UGF can be implemented easily using the Excel software. It is very easy to use Excel and one can easily determine the performance of the MSS using the Excel software. The authors explained the detailed procedure that how this can be implemented on the Excel software. For that, we took one example from the paper of Ding and Lisnianski [21]. Authors compared the results, they found that their results are exactly the same as the results obtained by Ding and Lisnianski [21] in their paper without any error. Thus, we hope that this research will be of great help to the research community as it can save their valuable time.

1.8   FUTURE SCOPE

The approach shown in this chapter for the implementation of the UGF technique is very easy and one can easily obtain all the system states without any difficulty. One can also use the same procedure for the k-out-of-n system. Further extension of the work is also possible if one tries to implement consecutive k-out-of-n system using Excel software.

CONFLICT OF INTEREST

The authors confirm that there is no conflict of interest to declare for this publication.

REFERENCES

1. Ushakov, I. A. (1986). A universal generating function. *Soviet Journal of Computer and Systems Sciences, 24*(5), 118–129.
2. Levitin, G. (2005). *The universal generating function in reliability analysis and optimization* Vol. 6. London: Springer.
3. Levitin, G., & Lisnianski, A. (1999). Importance and sensitivity analysis of multi-state systems using the universal generating function method. *Reliability Engineering & System Safety, 65*(3), 271–282.

4. Lisnianski, A., Levitin, G., Ben-Haim, H., & Elmakis, D. (1996). Power system structure optimization subject to reliability constraints. *Electric Power Systems Research*, *39*(2), 145–152.
5. Wen, K., Li, Y., Yang, Y. A. N. G., & Gong, J. (2018). Reliability evaluation of compressor systems based on universal generating function method. *Journal of Shanghai Jiaotong University (Science)*, *23*(2), 291–296.
6. Levitin, G., & Xing, L. (2010). Reliability and performance of multi-state systems with propagated failures having selective effect. *Reliability Engineering & System Safety*, *95*(6), 655–661.
7. Chaturvedi, S. K., Basha, S. H., Amari, S. V., & Zuo, M. J. (2012). Reliability analysis of generalized multi-state k-out-of-n systems. *Proceedings of the Institution of Mechanical Engineers, Part O: Journal of Risk and Reliability*, *226*(3), 327–336.
8. Yingkui, G., & Jing, L. (2012). Multi-state system reliability: A new and systematic review. *Procedia Engineering*, *29*, 531–536.
9. Lisnianski, A. (2012). L z-Transform for a discrete-state continuous-time Markov process and its applications to multi-state system reliability. In *Recent advances in system reliability* (pp. 79–95). London: Springer.
10. Sallak, M., Schön, W., & Aguirre, F. (2013). Reliability assessment for multi-state systems under uncertainties based on the Dempster–Shafer theory. *IIE Transactions*, *45*(9), 995–1007.
11. Meenakshi, K., & Singh, S. B. (2016). Availability assessment of multi-state system by hybrid universal generating function and probability intervals. *International Journal of Performability Engineering*, *12*(4), 321–339.
12. Meenakshi, K., & Singh, S. B. (2018). Reliability analysis of ((e, f), k, lc)/(m, n): F system under multiple failure using universal generating function. *International Journal of Reliability and Safety*, *12*(3), 292–305.
13. Meena, K. S., & Vasanthi, T. (2016). Reliability analysis of mobile ad hoc networks using universal generating function. *Quality and Reliability Engineering International*, *32*(1), 111–122.
14. Nair, V. G., & Manoharan, M. (2018). Dynamic multi state system reliability analysis of power generating system using Lz-transformation. *ProbStat Forum*, *11*, 81–90.
15. Levitin, G. (2011). Reliability of multi-state systems with common bus performance sharing. *IIE Transactions*, *43*(7), 518–524.
16. Wang, G., Duan, F., & Zhou, Y. (2018). Reliability evaluation of multi-state series systems with performance sharing. *Reliability Engineering & System Safety*, *173*, 58–63.
17. Dong, W., Liu, S., Fang, Z., & Cao, Y. (2019). Reliability evaluation of uncertain multi-state systems based on weighted universal generating function. *International Journal of Performability Engineering*, *15*(1), 167–178.
18. Lisnianski, A., Frenkel, I., & Karagrigoriou, A. (Eds.). (2017). *Recent advances in multi-state systems reliability: Theory and applications.* Springer, Switzerland.
19. Percy, D. F. (2013). *Recent advances in system reliability: Signatures, multistate systems and statistical inference.* Springer. https://onlinelibrary.wiley.com/doi/10.1002/9781118445112.stat03659
20. Kołowrocki, K. (2014). *Reliability of large systems.* Wiley StatsRef: Statistics reference online.
21. Ding, Y., & Lisnianski, A. (2008). Fuzzy universal generating functions for multi-state system reliability assessment. *Fuzzy Sets and Systems*, *159*(3), 307–324.

2 On the Optimization of the Cost Rate of Coherent Reliability Structures under Replacement Policy

Ioannis S. Triantafyllou
Department of Computer Science and Biomedical
Informatics, University of Thessaly, Lamia, Greece

CONTENTS

2.1 INTRODUCTION

Age replacement strategies have been broadly investigated in the field of Reliability Engineering and Statistical Modeling. Several age replacement scenarios have already been introduced and investigated in the literature. For some recent advances, the works provided by Eryilmaz and Ozkut (2020) or Safaei, Chatelet and Ahmadi (2020) are highly recommended.

In the present chapter, we consider the traditional age replacement policy model, under which the replacement takes place either at a pre-specified time *t* or at failure time (whichever comes first). Applying the aforementioned replacement strategy, we investigate the mean cost rate of coherent reliability structures under the so-called exponentiated Gamma distribution. More specifically, Section 2.2 provides the general framework of the computing procedure, which relies on the signature vector of the corresponding reliability structure. Some details and theoretical results referring to the underlying approach are also mentioned in order to make its implementation

DOI: 10.1201/9781003156291-2

more comprehensible. Section 2.3 offers some numerical results for illustrating the general approach, while some practical concluding remarks are also discussed. Finally, Section 2.4 summarizes the results provided in previous sections, while some hints for future research are also drawn.

2.2 NOTATIONS

In the present section, general notations are presented in order to make the remaining manuscript more readable. The main symbols that are utilized in the next sections are listed below.

i.i.d.: independent and identically distributed
MTTR: mean time to replacement
d_U: cost of an unpredictable replacement
d_P: cost of a predictable replacement
MCR: mean cost rate
s: signature vector
IFR: increasing failure rate
r(t): hazard rate
BS: bridge structure
$X_{i:n}$: ordered lifetime of a reliability system consisting of n components
E(X): first of moment of random variable X

2.3 A GENERAL APPROACH FOR DELIVERING THE MEAN COST RATE OF COHERENT STRUCTURES

In the present section, we study the cost behavior of coherent reliability systems under the common age replacement strategy. Let us next consider a coherent structure consisting of n independent and identically distributed (*i.i.d.*) components. Let T be the lifetime of the structure with continuous distribution F, while $X_1, X_2,...,X_n$ correspond to its components' lifetimes with common continuous distribution G and hazard rate function $r(t)$. The resulting structure operates under the standard age replacement strategy, namely the system is replaced either at fixed time t_0 ($t_0 > 0$) or at actual failure (see, e.g. Eryilmaz and Pekalp (2020)). In other words, it is plausible that the system could be replaced at time t_0 for deterrent reasons, while, in case of system's failure, its replacement is automatically activated (whatever comes first). Consequently, the average time to replacement is called *mean time to replacement* (*MTTR*, hereafter) and can be viewed as

$$MTTR = E(\min(T, t_0)) = \int_0^{t_0} (1 - F(t))dt, t > 0. \tag{2.1}$$

As it concerns the operating cost of the structure, there exist two separate types of it. The first one refers to the replacement cost at system's failure and expresses the cost of an unpredictable replacement (d_U, hereafter). The second type of operating cost corresponds to the replacement cost at fixed time t_0 and expresses the cost of a predictable replacement (d_P, hereafter). Having at hand the abovementioned design constants (d_U, d_P), we may readily define the so-called *mean cost rate* of the reliability structure (*MCR*, hereafter) as follows (see, e.g. Barlow and Proschan (1965))

$$MCR = \frac{d_U F(t_0) + d_P(1 - F(t_0))}{MTTR}, d_U > d_P. \tag{2.2}$$

Based on some theoretical results established by Samaniego (1985) or Eryilmaz and Pekalp (2020), they provided a necessary condition for the existence of a unique optimal design of a reliability system concerning its *MCR*.

Let us next denote by $\mathbf{s} = (s_1, s_2, ..., s_n)$ the signature vector of a coherent reliability structure with *i.i.d.* units, while

$$s_i = P(T = X_{i:n}), i = 1, 2, ..., n, \tag{2.3}$$

where $X_{1:n} \leq X_{2:n} \leq \cdots \leq X_{n:n}$ express the corresponding ordered random lifetimes (see, e.g. Navarro, Samaniego and Balakrishnan (2011) or Navarro et al. (2008)). Samaniego (1985) proved that, under the assumption that the components' lifetimes belong to the *increasing failure rate* (*IFR*) class, if the rational function

$$h(x) = \frac{\displaystyle\sum_{i=0}^{n-1}(n-i)s_i\binom{n}{i}x^i}{\displaystyle\sum_{i=0}^{n-1}\left(\sum_{j=i+1}^{n}s_j\right)\binom{n}{i}x^i}, \tag{2.4}$$

is increasing in $x \in (0, +\infty)$, then the system's lifetime T is also *IFR*. Denoting by n_1 the maximum number of failed components is shown such that the structure has not failed yet; Eryilmaz and Pekalp (2020) showed that if the following ensue

- $h(x)$ is increasing

- $(n - n_1)w\displaystyle\sum_{i=1}^{n}s_i E(X_{i:n}) > \frac{d_U}{d_U - d_P}$, where $w = \lim_{t \to \infty} r(t)$ \hfill (2.5)

then there exists a finite unique $t_1 > 0$ such that the *MCR* is minimized (see, also Zhao et al. (2017)).

Throughout the lines of the present section, a signature-based approach for optimizing the *MCR* of a reliability structure is described. More precisely, a step-by-step procedure for delivering the *MCR* of coherent structures consisting of *i.i.d.* components is presented below.

Step 1. Calculate the coordinates of the signature vector of the underlying structure.

Step 2. Study the monotonicity of function $h(x)$.

Step 3. Choose appropriately the design constants d_U and d_P.

Step 4. Determine the maximum number n_1 of failed components such that the structure can still operate.

Step 5. Specify the underlying distribution for components' lifetimes and derive the moments of the corresponding order statistics.

Step 6. Determine the unique value for which the mean cost rate is minimized.

Once we follow the abovementioned steps for a specific reliability structure, its mean cost rate under a predetermined distributional model would be at our hand. For implementing the step-by-step procedure presented above, the following remarks seem to be of some practical interest.

- The first step of the proposed procedure could be readily realized either based on the definition of signatures of coherent systems (see, e.g. Samaniego (1985)) or by implementing the generating function approach established by Triantafyllou and Koutras (2008). For some recent advances on the topic, the research works provided by Triantafyllou (2021) or Kumar, Tyagi and Ram (2020) are also recommended.
- The second step should be accomplished in order to activate the signature-based criterion for the *IFR* preservation under the formation of coherent structures, which has been proposed by Samaniego (1985).
- At the third step, the specific requirements or restrictions referring to the spending of the underlying problem should be taken into account. Note that d_1 is expected to be larger than the corresponding value of d_2.
- Steps 4 and 5 are related to the reliability system, which is under investigation, as well as depend on the underlying distributional features of its components.
- Based on the theoretical result proved by Eryilmaz and Pekalp (2020), the final step of the procedure leads the practitioner to the desired optimization.

We shall illustrate the abovementioned procedure for facing the relative optimization issue for well-known reliability structures in the next section of the present manuscript.

2.4   NUMERICAL RESULTS

In the present section, we proceed to numerical experimentation in order to illustrate the general framework described previously. We next investigate some well-known reliability structures along with traditional distributional assumptions for modeling their components' lifetimes. More precisely, we consider the bridge structure consisting of five components (see, e.g. Kumar, Pant and Ram (2017)) and the common r-within-consecutive-k-out-of-n: F system (see, e.g. Tong (1985) or Triantafyllou (2015)). For the abovementioned reliability structures, the exponentiated Gamma distribution is assumed for covering the behavior of their components.

2.4.1   THE BRIDGE STRUCTURE

In the present subsection, we consider the so-called bridge structure consisting of five i.i.d. components. An illustration of the abovementioned reliability system can be viewed in Figure 2.1.

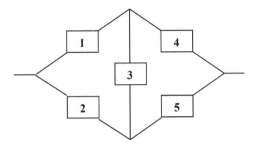

FIGURE 2.1   The bridge structure with $n = 5$ components.

It is widely known that a bridge structure with five i.i.d. components fails whenever the path from the one side of the system to the other is not feasible. For instance,
the failure of units 1, 2 or 4, 5 result in the total failure of the system. It is unsubtle
that there are more than two sets of failed components, which result in system's
overall failure. However, the failure of a single unit cannot cause bridge's failure.
Therefore, concerning its signature vector, it is obviously deduced that its first coordinate equals to zero. Following the definition of signature vector of a coherent reliability structure, we can readily obtain the remaining coordinates of the signature of
bridge with five i.i.d. components and therefore the first step of the proposed procedure would be accomplished. Indeed, we readily deduce that the probability that the
bridge fails upon the second and third ordered components' failure equals to 1/5 and
3/5, respectively. Moreover, given that the bridge structure cannot operate with only
one working component, it is immediate that the last coordinate of the corresponding signature vector coincides to zero and consequently the fourth one is equal to the
remaining part, which is equal to 1/5. To sum up, the signature vector of the bridge
structure is given by

$$s_{BS} = \left(0, \frac{1}{5}, \frac{3}{5}, \frac{1}{5}, 0\right). \tag{2.6}$$

Kindly note that the particular signature has been already investigated in detail and
studied in the literature (see, e.g. Samaniego (1985)). However, we chose to provide a brief description of its evaluation procedure for self-sufficiency of the present
manuscript.

Concerning the second step, we may rewrite $h(x)$ defined in Equation (2.4), for
the case of the bridge structure consisting of five i.i.d. components with common
distribution F as

$$h_{BS}(x) = \frac{\sum_{i=0}^{n-1}(n-i)s_{BS}(i)\binom{n}{i}x^i}{\sum_{i=0}^{n-1}\left(\sum_{j=i+1}^{n}s_{BS}(j)\right)\binom{n}{i}x^i} = \frac{2x(2x^2+9x+2)}{2x^3+8x^2+5x+1}. \tag{2.7}$$

Since the first derivative of h_{BS}, the abovementioned expression can be determined as

$$h'_{BS}(x) = \frac{4 + 2x(18 + x(35 - 2x(x - 6)))}{(1 + x(5 + 2x(4 + x)))^2}. \tag{2.8}$$

We easily confirm that its sign is positive for all $x > 0$. Consequently, h_{BS} is increasing in $x \in (0, +\infty)$ and based on Samaniego's result (1985); we conclude that the lifetime of the bridge structure is *IFR* whenever its components are independently distributed to a common *IFR* distribution. Additionally, the maximum number of failed components such that the bridge structure can still operate equals to 3, namely $n_1 = 3$.

In order to implement the step-by-step procedure mentioned in Section 2.2, we next assume that the components of the underlying reliability structure follow the exponentiated Gamma distribution with parameter $\theta > 0$. Generally speaking, Gupta et al. (1998) established a parametric approach for modeling failure time data under the assumption that a baseline continuous distribution function is available. Their exponentiated models include, among others, the so-called exponentiated Weibull distribution and the exponentiated Pareto distribution. We next assume that the components' lifetimes of the bridge structure follow the exponentiated Gamma distribution with parameter $\theta > 0$. Under the exponentiated Gamma model, the probability density function and the cumulative distribution function of components' lifetimes $X_1, X_2, ..., X_n$ are given as (see, e.g. Gupta et al. (1998))

$$f(t) = \theta t e^{-t} \left(1 - e^{-t} - t e^{-t}\right)^{\theta - 1}, t \geq 0 \text{ and } F(t) = \left(1 - e^{-t} - t e^{-t}\right)^{\theta}, t \geq 0, \tag{2.9}$$

respectively. Consequently, it is unambiguous that the corresponding hazard rate of $X_1, X_2, ..., X_n$ under the exponentiated Gamma distribution can be expressed as

$$r(t) = \frac{\theta t e^{-t} \left(1 - e^{-t} - t e^{-t}\right)^{\theta - 1}}{1 - \left(1 - e^{-t} - t e^{-t}\right)^{\theta}}, t \geq 0. \tag{2.10}$$

Gupta et al. (1998) proved also that the exponentiated Gamma distribution belongs to the *IFR* class for all $\theta > 1/2$, while it is readily deduced that $\lim_{t \to \infty} r(t) = 1$ for all real values of parameter θ. Before implementing the procedure described in Section 2.2, we should calculate the first moment of the corresponding order statistics $X_{1:n}, X_{2:n}, ..., X_{n:n}$, where $X_{1:n} < X_{2:n} < \cdots < X_{n:n}$. Shawky and Bakoban (2009) proved that the first moment of the ith order statistic $X_{i:n}$ obtained from the exponentiated distribution with positive integer-valued parameter θ is given as

$$E(X_{i:n}) = \frac{n!\theta}{(i-1)!(n-i)!} \sum_{m=0}^{n-i} \sum_{j=0}^{\theta(i+m)-1} \sum_{y=0}^{j} (-1)^{m+j} \tag{2.11}$$

$$\times \binom{n-i}{m} \binom{\theta(i+m)-1}{j} \binom{j}{y} \frac{(y+2)!}{(j+1)^{y+3}}.$$

To sum up, the bridge structure with five i.i.d. components with a common exponentiated Gamma distribution with parameter θ satisfies the following:

- The components' lifetimes are *IFR* for $\vartheta > 1/2$.
- The rational function $h_{BS}(x)$ is increasing in $x \in (0,+\infty)$.
- The system's lifetime T is *IFR* (based on the two previously mentioned bullets).
- The maximum number of failed components such that the bridge structure can still operate equals to $n_1 = 3$.
- $w = \lim\limits_{t \to \infty} r(t) = 1$.
- The signature vector of the bridge structure is given as $s_{BS} = \left(0, \dfrac{1}{5}, \dfrac{3}{5}, \dfrac{1}{5}, 0\right)$.

We are now ready to calculate the left-hand side of inequality (Equation 2.5) for the bridge structure. Indeed, we easily deduce that

$$(n - n_1)w \sum_{i=1}^{n} s_i E(X_{i:n}) = 2 \cdot \left(\frac{1}{5} \cdot E(X_{2:5}) + \frac{3}{5} \cdot E(X_{3:5}) + \frac{1}{5} \cdot E(X_{4:5})\right). \qquad (2.12)$$

Recalling formula (2.11), we next assume that parameter θ = 2 and the following ensue

$$E(X_{2:5}) = 1.94237, E(X_{3:5}) = 2.55207, E(X_{4:5}) = 3.3111.$$

Consequently, the expression given in Equation (2.12) equals to 5.16387. Based on Equation (2.5), once we determine properly the parameters d_U and d_P such that $d_U/(d_U - d_P) < 5.16387$, we readily conclude that there exists a finite unique $t_1 > 0$ such that the *MCR* of the bridge structure is minimized. We next determine the optimal replacement time for the bridge structure with five exponentiated Gamma distributed components for several values of the design parameters. The corresponding numerical results are displayed in Table 2.1.

Based on the numerical results displayed in Table 2.1, we readily conclude that

- The *MCR* decreases as the *parameter d_U* decreases (under the assumption that d_P does not change).
- The *optimal replacement time* decreases as the *parameter d_U* decreases (under the assumption that d_P does not change).
- The *MCR* decreases as the *parameter d_P* decreases (under the assumption that d_U does not change).
- The *optimal replacement time* increases as the *parameter d_P* decreases (under the assumption that d_U does not change).

2.4.2 THE *R*-WITHIN-CONSECUTIVE-*K*-OUT-OF-*N*: *F* SYSTEM

In the present subsection, we consider the so-called *r*-within-consecutive *k*-out-of-*n*: *F* structure, which was established by Tong (1985), and since then it has attracted

TABLE 2.1
The Minimized *MCR* and the Optimal Replacement Time for the Bridge Structure

Design Parameters (d_U, d_p)	Optimal Replacement Time (t_1)	MCR
(8,2)	8.50864×10^{21}	9.4022×10^{-22}
(8,1.5)	8.97752×10^{21}	8.91114×10^{-22}
(8,1)	1.06375×10^{22}	7.52056×10^{-22}
(7,2)	7.89953×10^{21}	8.86128×10^{-22}
(7,1.5)	1.0094×10^{22}	6.93479×10^{-22}
(7,1)	1.41864×10^{22}	4.9343×10^{-22}
(6,2)	7.37285×10^{21}	8.13797×10^{-22}
(6,1.5)	9.38675×10^{21}	6.39199×10^{-22}
(6,1)	1.24662×10^{22}	4.813×10^{-22}
(5,2)	7.22952×10^{21}	6.91609×10^{-22}
(5,1.5)	9.26685×10^{21}	5.39557×10^{-22}
(5,1)	1.27608×10^{22}	3.91825×10^{-22}
(4,2)	6.48742×10^{21}	6.16578×10^{-22}
(4,1.5)	7.60539×10^{21}	5.25943×10^{-22}
(4,1)	8.08079×10^{21}	4.95001×10^{-22}

Underlying distribution: exponentiated Gamma distribution with parameter $\theta = 2$.

some research interest (see, e.g. Triantafyllou and Koutras (2011) or Eryilmaz, Kan and Akici (2009)). Generally speaking, the system consists of n components and fails if, and only if, there exist k consecutive units that include among them at least r failed components.

For illustration purposes, let us next depict the so-called r-within-consecutive k-out-of-n: F structure with the aid of a diagram. Figure 2.2 illustrates some failure scenarios (not all of them) for the r-within-consecutive k-out-of-n: F structure for predetermined cases concerning its design parameters. Note that a gray-filled box indicates a failed unit, while a blank box indicates a component that still operates.

Each one of the above scenarios results in the overall failure of the underlying structure. Indeed, one may readily observe that if any of the abovementioned schemes occurs, the failure criterion of the 2-within-consecutive-3-out-of-9: F structure is met.

In order to implement the step-by-step procedure mentioned in Section 2.2, we next assume that the components' lifetimes of the underlying reliability structure follow the exponentiated Gamma distribution mentioned before (see, e.g. Gupta et al. (1998)). Based on formulae (2.7–2.9) provided in the previous subsection of the present manuscript, we shall next deliver some numerical results concerning the optimization issue of *MCR* of the 2-within-consecutive 3-out-of-9: F system.

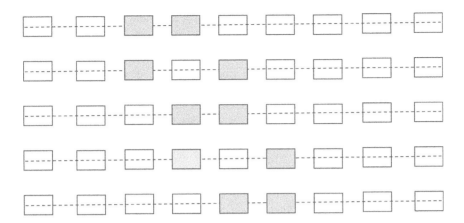

FIGURE 2.2   Failure scenarios for the 2-within-consecutive 3-out-of-9: F system.

Following a parallel argumentation as the one presented previously for the bridge structure, we may easily conclude that, for 2-within-consecutive 3-out-of-9: F system consisting of i.i.d. components with a common exponentiated Gamma distribution, the following holds true.

- The signature vector of the 2-within-consecutive 3-out-of-9: F system is given as $s_{within} = \left(0, \dfrac{5}{12}, \dfrac{13}{28}, \dfrac{5}{42}, 0,0,0,0,0\right)$ (for more details, see Triantafyllou and Koutras (2011)).
- The components' lifetimes are *IFR* for $\vartheta > 1/2$.
- The rational function defined in Equation (2.4) can be expressed for the 2-within-consecutive 3-out-of-9: F system as

$$h_{within}(x) = \frac{60x^3 + 117x^2 + 30x}{10x^3 + 21x^2 + 9x + 1}. \tag{2.13}$$

- The function $h_{within}(x)$ given in Equation (2.11) is increasing in $x \in (0, +\infty)$.
- The system's lifetime T is *IFR* (based on the three previously mentioned bullets).
- The maximum number of failed components such that the 2-within-consecutive 3-out-of-9: F system can still operate equals to $n_1 = 3$.
- $w = \lim\limits_{t \to \infty} r(t) = 1$.

We are now ready to calculate the left-hand side of inequality (Equation 2.5) for the 2-within-consecutive 3-out-of-9: F system. Indeed, we easily deduce that

$$(n - n_1)w \sum_{i=1}^{n} s_i E(X_{i:n}) = 6 \cdot \left(\frac{5}{12} \cdot E(X_{2:9}) + \frac{13}{28} \cdot E(X_{3:9}) + \frac{5}{42} \cdot E(X_{4:9}) \right). \tag{2.14}$$

Recalling formula (2.9), we assume once again that $\theta = 2$ and the following ensue

$$E(X_{2:9}) = 1.47896, \ E(X_{3:9}) = 1.82836, \ E(X_{4:9}) = 2.16536.$$

Consequently, the expression given in Equation (2.14) equals to 10.3374. Based on Equation (2.5), once we determine properly the parameters d_U and d_P such that $d_U/(d_U - d_P) < 10.3374$, we readily conclude that there exists a finite unique $t_1 > 0$ such that the MCR of the 2-within-consecutive 3-out-of-9: F system is minimized. We next determine the optimal replacement time for the 2-within-consecutive 3-out-of-9: F system with exponentiated Gamma distributed components for several values of the design parameters. The corresponding numerical results are displayed in Table 2.2.

Based on the numerical results displayed in Table 2.2, we readily conclude that

- The MCR decreases as the *parameter d_U* decreases (under the assumption that d_P does not change).
- The *optimal replacement time* decreases as the *parameter d_U* decreases (under the assumption that d_P does not change).
- The MCR decreases as the *parameter d_P* decreases (under the assumption that d_U does not change).

TABLE 2.2
The Minimized MCR and the Optimal Replacement Time for the 2-within-Consecutive 3-out-of-9: F System

Design Parameters (d_U, d_P)	Optimal Replacement Time (t_1)	MCR
(8,2)	1.0297×10^{12}	7.76922×10^{-12}
(8,1.5)	1.05717×10^{12}	7.5674×10^{-12}
(8,1)	1.10298×10^{12}	7.39846×10^{-12}
(7,2)	1.01583×10^{12}	6.8909×10^{-12}
(7,1.5)	1.04482×10^{12}	6.6997×10^{-12}
(7,1)	1.08022×10^{12}	6.48016×10^{-12}
(6,2)	9.98898×10^{11}	6.00662×10^{-12}
(6,1.5)	1.01143×10^{12}	5.88723×10^{-12}
(6,1)	1.06749×10^{12}	5.62067×10^{-12}
(5,2)	9.77562×10^{11}	5.11477×10^{-12}
(5,1.5)	1.01058×10^{12}	4.94765×10^{-12}
(5,1)	1.05129×10^{12}	4.75608×10^{-12}
(4,2)	9.49481×10^{11}	4.21283×10^{-12}
(4,1.5)	1.01015×10^{12}	4.03134×10^{-12}
(4,1)	1.02971×10^{12}	3.88461×10^{-12}

Underlying distribution: exponentiated Gamma distribution with parameter $\theta = 2$.

2.5   DISCUSSION

In the present chapter, the traditional age replacement policy strategy is applied to two different well-known coherent structures with *i.i.d.* components, in order to shed light on their mean cost rate. For illustration purposes, the so-called exponentiated Gamma distribution is considered to be the distribution of the corresponding components' lifetimes. It is remarkable that, for both systems considered, the optimal replacement time does not seem to come at an early operating stage, while the proper choices of the design parameters of the replacement policy strategy could lead to the desirable cost rate. It seems to be of some interest to investigate the cost behavior of additional members of the so-called consecutive-type reliability structures and/or by the aid of different cost measures.

REFERENCES

Barlow, R. E. & Proschan, F. (1965). *Mathematical Theory of Reliability*, John Wiley & Sons, New York, NY.

Eryilmaz, S., Kan, M. & Akici, F. (2009). Consecutive-*k*-within-*m*-out-of-*n: F* system with exchangeable components, *Naval Research Logistics*, **56**, 503–510.

Eryilmaz, S. & Ozkut, M. (2020). Optimization problems for a parallel system with multiple types of dependent components, *Reliability Engineering & System Safety*, **199**, 106911.

Eryilmaz, S. & Pekalp, M. H. (2020). On optimal age replacement policy for a class of coherent systems, *Journal of Computational and Applied Mathematics*, **377**, 112888.

Gupta, R. C., Gupta, R. D. & Gupta, P. L. (1998). Modeling failure time data by Lehmann alternatives, *Communications in Statistics: Theory & Methods*, **27**, 887–904.

Kumar, A., Pant, S. & Ram, M. (2017). System reliability optimization using gray wolf optimizer algorithm, *Quality and Reliability Engineering International*, **33**, 1327–1335.

Kumar, A., Tyagi, S. & Ram, M. (2020). Signature of bridge structure using universal generating function, *International Journal of System Engineering and Management*, **12**, 53–57.

Navarro, J., Samaniego, F. J. & Balakrishnan, N. (2011). Signature-based representations for the reliability of systems with heterogeneous components, *Journal of Applied Probability*, **48**, 856–867.

Navarro, J., Samaniego, F. J., Balakrishnan, N. & Bhattacharya, D. (2008). On the application and extension of system signatures in engineering reliability, *Naval Research Logistics*, **55**, 313–327.

Safaei, F., Chatelet, E. & Ahmadi, J. (2020). Optimal age replacement policy for parallel and series systems with dependent components, *Reliability Engineering & System Safety*, **197**, 106798.

Samaniego, F. J. (1985). On closure of the IFR class under formation of coherent systems, *IEEE Transactions on Reliability*, **34**, 69–72.

Shawky, A. I. & Bakoban, R. A. (2009). Order statistics from exponentiated gamma distribution and associated inference, *International Journal of Contemporary Mathematical Sciences*, **4**, 71–91.

Tong, Y. L. (1985). A rearrangement inequality for the longest run, with an application to network reliability, *Journal of Applied Probability*, **22**, 386–393.

Triantafyllou, I. S. (2015). Consecutive-type reliability systems: an overview and some applications, *Journal of Quality and Reliability Engineering*, **2015**, 20 pages. Article ID 212303.

Triantafyllou, I. S. (2021). Reliability study of <*n,f,*2> systems: a generating function approach, *International Journal of Mathematical, Engineering and Management Sciences*, **6**, 44–65.

Triantafyllou, I. S. and Koutras, M. V. (2008). On the signature of coherent systems and applications, *Probability in the Engineering and Informational Science*, **22**, 19–35.

Triantafyllou, I. S. & Koutras, M. V. (2011). Signature and *IFR* preservation of 2-within-consecutive-k-out-of-n: *F* systems and applications, *IEEE Transactions on Reliability*, **60**, 315–322.

Zhao, X., Al-Khalifa, K. N., Hamouda, A. M. & Nakagawa, T. (2017). Age replacement models: a summary with new perspectives and methods, *Reliability Engineering & System Safety*, **161**, 95–105.

3 The Disjoint Minimal Paths Reliability Problem

Majid Forghani-elahabad
Center of Mathematics, Computing, and Cognition,
Federal University of ABC, Santo André, SP, Brazil

CONTENTS

DOI: 10.1201/9781003156291-3

3.1   INTRODUCTION

The shortest path problem is very well-known and one of the most attractive problems in the operations research area due to many applications in practice and theory [1]. Several variants of this problem, including the shortest delay path, the least cost path, the largest capacity path, and the quickest path, have been studied in the literature [2–4]. The quickest path problem, which is the more related variant to this chapter, is to determine a path for transmitting some given amount of commodities (flow or data) from a source to a destination with the minimum transmission time [2]. Although the problem was initiated by Moore [5], the notion and the first algorithm were proposed by Chen and Chin [2]. Since then, many algorithms have been developed to address this problem in the literature [6, 7]. In the quickest path problem, the arcs' capacities are assumed to be deterministic. However, due to operations activities, failures, maintenance needs, and so forth, the arcs' capacities in many real-world systems may vary [3, 8–12]. Such networks in which the arcs and maybe the nodes have stochastic capacities with more than two states are called stochastic-flow networks or multistate flow networks (MFNs) [13–17]. Extending the quickest path problem to the MFNs, Lin [3] introduced the quickest path reliability problem in which the probability of sending d units of commodities (data or flow) from a source node to a destination node through a single minimal path (MP) in the network within T units of time is calculated. Later, the author extended the problem to the cases with two or more disjoint MPs (DMPs) [18, 19]. Since then, several algorithms have been proposed in the literature for these problems [9–11, 20–22]. These problems aim at assessing the reliability of MFNs under some time limits. Without consideration of time limit, many algorithms have been proposed in the literature to compute the network reliability based on minimal cuts [8, 13, 14, 23–28] or MPs [12, 16, 17, 23, 29–38]. However, as the time limit is of great importance in many real-world systems, this chapter focuses on the quickest path reliability problem and its extensions.

In this chapter, the required notations, nomenclature, and assumptions are stated in Section 3.2. We provide some preliminaries in Section 3.3. The system reliability is discussed in Section 3.4. Section 3.5 studies the quickest path reliability problem. The extensions of the problem to the cases of two or more DMPs are individually discussed in Sections 3.6 and 3.7. For each problem, an algorithm is stated and illustrated through a benchmark network example. Moreover, each section provides the complexity results of the corresponding algorithm. Afterward, we introduce and address some related optimization problems in Section 3.8. Finally, Section 3.9 concludes the chapter.

3.2 ACRONYMS, NOTATIONS, NOMENCLATURE, AND ASSUMPTIONS

3.2.1 ACRONYMS

MFN	Multistate flow network
MP	Minimal path
DMP	Disjoint minimal path
SSV	System state vector

3.2.2 NOTATIONS

m, n the number of arcs and nodes in the network, respectively.

N $N = \{1, 2, \ldots, n\}$ is the set of nodes in which nodes 1 and n are the source and the sink nodes.

A $A = \{a_i \mid 1 \le i \le m\}$ is the set of arcs.

L $L = (l_1, l_2, \ldots, l_m)$ is a vector where l_i denotes the lead time of arc a_i, for $i = 1, \ldots, m$.

M $M = (M_1, M_2, \ldots, M_m)$ is a vector where M_i denotes the maximum capacity of arc a_i, for $i = 1, \ldots, m$.

G $G = G(N, A, L, M)$ is a multistate flow network with the set of nodes N, set of arcs A, lead time vector of L, and maximum capacity vector of M.

τ the transmission time limit.

x_i current capacity of arc a_i that takes value randomly from $\{0, 1, \ldots, M_i\}$ according to a given probability distribution function, for $i = 1, \ldots, m$.

X $X = (x_1, x_2, \ldots, x_m)$ is the current SSV.

q number of all the MPs in the network.

k number of the given DMPs in the network.

P_j jth MP, for $j = 1, 2, \ldots, q$.

LP_j lead time of P_j, for $j = 1, 2, \ldots, q$.

$CP_j(X)$ capacity of P_j under SSV of X, for $j = 1, 2, \ldots, q$.

d a demand value which is a nonnegative integer number and shows the amount of a commodity which should be transmitted from a source node to a sink node.

D $D = (d_1, d_2, \ldots, d_k)$ is a *policy vector*, which satisfies $d_1 + d_2 + \cdots + d_k = d$, with d_j denoting the amount of commodity which should be sent through P_j, for $j = 1, 2, \ldots, k$.

$h(P, d, X)$ the transmission time to send d units of commodity through minimal path p under SSV of X.

$h(d, X)$ the transmission time for sending d units of a commodity from node 1 to node n through the network under SSV of X.

$h^1(d, X)$ the transmission time to send d units of commodity from node 1 to node n through a single MP under SSV of X.

$h^k(D,X)$ the transmission time to send $d = d_1 + \cdots + d_k$ units of commodity from node 1 to node n through k DMPs according to the *policy vector* of $D = (d_1,\ldots,d_k)$ under SSV of X (see more details in Section 3.6 for the case of $k = 2$ and in Section 3.7 for the general case).

$R^{(1)}_{(d,\tau)}$ the network reliability when the commodity is transmitted through a single MP (more details in Section 3.5).

$R^{(2)}_{(d,\tau)}$ the network reliability when the commodity is transmitted through two DMPs.

$R^{(k)}_{(d,\tau)}$ the network reliability when the commodity is transmitted through k DMPs where $k \geq 2$.

(d,τ)-MP^1 see Definitions 3.1 and 3.2.

(d,τ)-MP^2 see Definitions 3.3 and 3.4.

(d,τ)-MP^k see Definitions 3.5 and 3.6.

σ the number of final solutions with no duplicates in each problem, that is the number of (d,τ)-MP^1 s in the quickest path reliability problem, (d,τ)-MP^2 s in the two DMPs reliability problem and (d,τ)-MP^k s in the several DMPs reliability problem.

δ an upper bound for the number of *policy vector* in each problem.

3.2.3 NOMENCLATURE

- For two SSVs $X = (x_1, x_2, \ldots, x_m)$ and $Y = (y_1, y_2, \ldots, y_m)$, we say that $X \leq Y$ if $x_i \leq y_i$, for any $i = 1, 2, \ldots, m$. And $X < Y$ when $X \leq Y$ and there exists at least one $1 \leq j \leq m$ such that $x_j < y_j$.

- An SSV, say $X \in \Psi$, is said to be a minimal vector if there is no $Y \in \Psi$ such that $Y < X$. It is noted that for X to be a minimal vector in Ψ, it does not need to be less than all the other vectors, but it is only required that there is no other vector less than it. For instance, all the vectors in $\{(2,3,1),(1,2,3),(3,2,1)\}$ are the minimal vectors.

- A path is a set of adjacent arcs to send commodities from the source node to the destination node.

- A MP is a path with no proper subsets being a path. For example, $P_1 = \{a_1, a_4, a_7\}$ is an MP for the network in Figure 3.1, but $P = \{a_1, a_4, a_2, a_3, a_8\}$ is not because its proper subset, $P' = \{a_3, a_8\}$, is still a path.

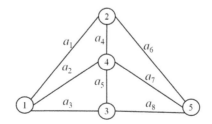

FIGURE 3.1   An adjusted benchmark network example taken from Ref. [25].

3.2.4 ASSUMPTIONS

Here, we state some general assumptions that are considered throughout this chapter. However, as several problems are discussed in this chapter, some particular assumptions related to each problem are stated in the corresponding sections.

1. Each node is perfectly reliable, i.e., deterministic.
2. The capacity of arc $a_i \in A$ is stochastic with a given probability distribution which takes integer values from $\{0,1,...,M_i\}$, for $i = 1,2,...,m$.
3. Capacity of each arc is s-independent from the one for any other arc.
4. Flow in the network satisfies the flow conservation law.
5. The required MPs are provided in advance.

3.3 PRELIMINARIES

The current capacity of arc a_i, namely x_i, shows the maximum amount of commodities (flow or data) which can be sent through arc a_i per unit of time, and l_i shows the required time to transmit one to x_i units of a commodity through arc a_i, for $i = 1,2,...,m$. Now, the transmission time to send $d > x_i$ units of a commodity through arc a_i is equal to

$$l_i + \left\lceil \frac{d}{x_i} \right\rceil,\tag{3.1}$$

where $[\alpha]$ is the smallest integer number greater than or equal to α. It is noted that the lead time of a_i being l_i means that one needs l_i units of time to send $1,2,...,x_i$ units of a commodity from the initial node to the terminal node of a_i, that is, the required time for sending one, two or till x_i units of a commodity is the same, and this is why $\left\lceil \frac{d}{x_i} \right\rceil$ is considered in Equation (3.1).

For instance, in Figure 3.1, a network with $N = \{1,2,3,4,5\}$ and $A = \{a_1,a_2,...,a_8\}$ is given. One can consider $M = (5,3,4,3,2,4,5,3)$ and $L = (2,2,3,1,2,2,3,1)$ as the max-capacity and lead time vectors. This way, a current system state vector (SSV) for the network can be a 9-tuple nonnegative integer-valued vector $X = (x_1,...,x_9) \leq M$; for example, $X^* = (3,3,3,2,2,4,3,3)$. One observes that under X^* at most three units of a commodity can be transmitted from node 1 to node 2 through a_1 per unit of time, and as $l_1 = 2$, it takes two units of time to transmit these three units of the commodity through this arc. One notes that sending 10 units of the commodity from node 1 to node 2 through a_1 under X^* takes $2 + [10/3] = 6$ units of time. In fact, starting the transmission from node 1, $x_1 = 3$ units of the commodity can be pumped into the arc per unit of time and it takes two units of time for the first commodity to arrive at node 2. After that, $x_1 = 3$ units of the commodity are pumped out from node 2 per unit of time, and hence since then it takes $[10/3] = 4$ units of time until all of the commodity arrives at the destination node. Therefore, the total required time is $2 + [10/3] = 6$.

A path is a set of adjacent arcs through which one can send commodities from the source node to the destination node. For instance, the set of $P_1 = \{a_1,a_4,a_7\}$ is a path from node 1 to node 5 in the network of Figure 3.1. Let $P_j = \left\{ a_{j_1}, a_{j_2}, \ldots, a_{j_{nj}} \right\}$ be a path. Then, its lead time, denoted by LP_j, is equal to

$$LP_j = \sum_{r=1}^{nj} l_{j_r}. \tag{3.2}$$

Also, its capacity under SSV of X, denoted by $CP_j(X)$, is equal to

$$CP_j(X) = \min\left\{ x_{j_1}, x_{j_2}, \ldots, x_{j_{nj}} \right\}. \tag{3.3}$$

For instance, the lead time of path $P_1 = \{a_1, a_4, a_7\}$ in Figure 3.1 is equal to $LP_1 = l_1 + l_4 + l_7 = 2 + 1 + 3 = 6$ and its capacity under $X^* = (3,3,3,2,2,4,3,3)$ is equal to $CP_1(X) = \min\{3,2,3\} = 2$.

Now, let $h(P_j, d, X)$ be the transmission time for sending d units of a commodity from node 1 to node n through path $P_j = \left\{ a_{j_1}, a_{j_2}, \ldots, a_{j_{nj}} \right\}$ under SSV of X. Then, we have

$$h(P_j, d, X) = LP_j + \left\lceil \frac{d}{CP_j(X)} \right\rceil. \tag{3.4}$$

For example, the required time to transmit 10 units of a commodity from node 1 to node 5 through $P_1 = \{a_1, a_4, a_7\}$ in Figure 3.1 under SSV of $X^* = (3,3,3,2,2,4,3,3)$ is equal to $h(P_1, 10, X) = LP_1 + \left\lceil \frac{10}{CP_1(X)} \right\rceil = 6 + [10/2] = 11$.

Hence, for a given d amount of commodity, an elementary approach is to determine all the paths and then calculate the transmission time for sending d units of a commodity through each path to choose the quickest path. Several efficient approaches have been proposed in the literature for a deterministic network flow [2, 6, 7] to which the reader may refer for more details. However, a MFN situation is different as the arcs' (and maybe nodes') capacities are not fixed and may vary. The extension of the quickest path problem to the MFNs is called the quickest path *reliability* problem [3] on which the focus of this chapter is. In the following sections, we first talk briefly about the system reliability and then discuss the quickest path reliability problem and its generalization to the cases with two or more DMPs.

3.4   SYSTEM RELIABILITY

Assume that d units of a commodity are given to transmit from the source node 1 to destination node n within a given τ units of time. The system reliability, denoted by $R_{(d,\tau)}$ for the general case, is the probability of success in reaching this goal. Let $h(d, X)$ be the transmission time for sending d units of a commodity from node 1 to node n through the network under SSV of X and $\Psi_d = \{0 \le X \le M \mid h(d, X) \le \tau\}$. Then, one can see that $R_{(d,\tau)} = \Pr\{X \mid X \in \Psi_d\}$. Now, one may use the following lemma to make the calculation simpler.

LEMMA 3.1

Let X and Y be two SSVs. If $X \leq Y$, then $h(d,X) \geq h(d,Y)$.

Proof. Let P_j be an arbitrary path in the network. As $X \leq Y$, we have $CP_j(X) \leq CP_j(Y)$, and hence according to Equation (3.4), we have $h(P_j, d, X) \geq h(P_j, d, Y)$. Consequently, $h(d,X) \geq h(d,Y)$ as P_j is an arbitrary path.

Letting $\Psi_{d,\min} = \{X^1, X^2, \ldots, X^\sigma\}$ be the set of all the minimal vectors in Ψ_d and $A_r = \{X \mid X \geq X^r\}$, for $r = 1, 2, \ldots, \sigma$. It is vivid that $\Psi_d = \underset{1 \leq r \leq \lambda}{\cup} A_r$, and consequently, the system reliability can be computed by using the sum of disjoint products as follows [39]. Let $B_1 = A_1, B_i = A_i - \overset{i-1}{\underset{r=1}{\cup}} A_r$, for $i = 2, 3, \ldots, \sigma$ to have the disjoint sets with the same union. As a result, we have

$$R_{(d,\tau)} = \sum_{r=1}^{\lambda} \Pr(B_r), \qquad (3.5)$$

where $\Pr(B_r) = \sum_{X \in B_r} \Pr(X)$, and $\Pr(X) = \prod_{j=1}^{m} \Pr(x_j)$.

A notable point (or challenge) to determine the set Ψ_d is that $h(d,X)$ varies depending on the paths through which the d units of a commodity are transmitted. For example, the transmission time under X can be lower when using two disjoint paths than when using only a single path. One may define $h(d,X)$ as the minimum possible transmission time to complete the job under X. However, in this case, calculating $h(d,X)$ is complicated and somehow an optimization problem by itself. Therefore, in this chapter, we restrict our attention to the cases in which the commodity is transmitted through a single path, two disjoint paths, and more than two disjoint paths. To this end, the definition of system reliability is adopted for each case in the corresponding section.

3.5 THE QUICKEST PATH RELIABILITY PROBLEM

Extending the quickest path problem to system reliability, Lin [3] introduced the quickest path reliability problem that is discussed in this section. The central assumption in this problem is that the whole commodity should be transmitted through only a single path. Let $h^1(d,X)$ be the transmission time for sending d units of a commodity under X through only one path. Let also $P_1, P_2, \ldots, P_q$ be all the MPs in the network. Hence, q is the number of all the MPs in the network. We note that for any path, say P, in the network, there is at least one $j = 1, 2, \ldots q$ such that $P_j \subset P$. If P is itself an MP, there is nothing to prove. Otherwise, it has a proper subset which is a path. If this subset is not an MP, one may repeat the argument to arrive at a subset that is an MP.

PROPOSITION 3.1

Let P and P' be two paths in a network. If $P \subset P'$, then for any system state vector (SSV) of X and any given demand of d, we have $h(P,d,X) \le h(P',d,X)$.

Proof. As $P \subset P'$, the lead time of P is vividly less than or equal to the lead time of P'. Also, it can be seen that as the number of arcs in a path increases, its capacity may decrease. Hence, the capacity of P is more than or equal to P'. Thus, the proof is complete according to Equation (3.4).

Therefore, the quickest path is one of the MPs in the network, and accordingly one can calculate $h^1(d,X)$ as follows.

$$h^1(d,X) = \min\{h(P_j,d,X) \mid j = 1,2,\ldots,q\}. \tag{3.6}$$

The adopted definition for the system reliability for this case, denoted by $R^{(1)}_{(d,\tau)}$, is defined as the probability of transmitting d units of a commodity from the source to the destination within time τ through only one MP. Hence, we have $R^{(1)}_{(d,\tau)} = \Pr\{X \mid h^1(d,X) \le \tau\}$. The quickest path reliability problem is to compute such a probability. To this end, one needs to determine all the X s with $h^1(d,X) \le \tau$. Let first see the following definitions that are adopted from the definitions of similar concepts in the literature [3, 9, 10].

DEFINITION 3.1

An SSV of X is called (d,τ)-MP^1 candidate, if d units of a commodity can be transmitted from the source to the sink through only one MP under X within time τ.

DEFINITION 3.2

Assuming that Ψ_d is the set of all the (d,τ)-MP^1 candidates, then any $X \in \Psi_{d,\min}$ is called a (d,τ)-MP^1.

According to our discussion in Section 3.4, one observes that $R^{(1)}_{(d,\tau)}$ can be computed through Equation (3.5) by having all the (d,τ)-MP^1 s at hand. Hence, we turn our attention to the determination of all the (d,τ)-MP^1 s. The following lemma proven in Ref. [3] gives us an equivalent definition for (d,τ)-MP^1 s.

LEMMA 3.2

An SSV of X is a (d,τ)-MP^1 if and only if (1) $h^1(d,X) \le \tau$ and (2) $h^1(d,Y) \ge \tau$ for any SSV, $Y \le X$.

If one sets the capacity of every arc in an MP, say P_r, to be positive and the capacity of all the other arcs to be zero, then the capacity of all the MPs except P_r will be zero. Accordingly, the transmission time under such SSV through every MP except P_r will be infinite. It is the key point to find all the (d,τ)-MP^1 s without determining the candidates. The following lemma shows how to determine the least capacity for every arc in an MP such that within τ units of time, one can transmit d units of a commodity through it.

LEMMA 3.3

Let P_j be an MP, the given time limit is greater than the path lead time, that is,

$$\tau > LP_j, \text{ and } c_j = \left\lceil \frac{d}{\tau - LP_j} \right\rceil. \text{ If } c_j \le CP_j(M), \text{ then the vector } X = (x_1, x_2, \dots, x_m)$$

defined below is the minimum SSV under which d units of a commodity can be sent through P_j within τ units of time.

$$\text{For } i = 1, 2, \dots, m, x_i = \begin{cases} x_i = c_j & \text{if } a_i \in P_j, x_i = 0 & \text{if } a_i \notin P_j. \end{cases} \tag{3.7}$$

Proof. It is easy to see that $c_j = \left\lceil \dfrac{d}{\tau - LP_j} \right\rceil$ is the smallest integer number such that $LP_j + \left\lceil \dfrac{d}{c_j} \right\rceil \le \tau$. Hence, as $c_j \le CP_j(M)$, it is the minimum possible capacity for P_j to have its transmission time less than or equal to τ. According to the definition of the capacity of an MP, it is clear that the above defined vector X is the minimal vector under which the capacity of P_j is equal to c_j, and consequently, the proof is complete.

It is noted that if $\tau \le LP_j$ for some MP, then it is impossible to send even one amount of commodity through this path within τ units of time. Moreover, if $\tau > LP_j$ but $c_j > CP_j(M)$ for some MP, the defined vector in Lemma 3.3 does not satisfy the condition $X \le M$. Therefore, in such cases, we do not have any (d,τ)-MP^1 associated with the considered MP. The following theorem proven in Ref. [3] shows that the determined vectors in Lemma 3.3 form all the (d,τ)-MP^1 s.

THEOREM 3.1

The vectors determined in Lemma 3.3 form the set of all the (d,τ)-MP^1 s in the network.

PROPOSITION 3.2

The number of (d,τ)-MP^1 s in a large enough network is less than or equal to MPs.

Proof. According to Lemma 3.3 and Theorem 3.1, there exists at most one (d,τ)-MP^1 corresponding to each MP, and hence the proof is complete.

3.5.1 AN ALGORITHM TO ADDRESS THE PROBLEM

Now, we are ready to propose the algorithm to address the quickest path reliability problem. Note that the following algorithm is a slightly improved version of the proposed algorithm in Ref. [3].

Algorithm 3.1 (network reliability evaluation based on (d,τ)-MP^{1}s)
Input: A multistate flow network with demand level d, time limit τ and all the MPs, say $P_1, P_2, \ldots, P_q$.
Output: The system reliability, that is $R_{(d,\tau)}^{(1)}$.
Step 1. Find the corresponding SSV with each MP, determined in Lemma 3.3, if it exists. Let $\{X^1, X^2, \ldots, X^\sigma\}$ be the set of all the obtained vectors (note that $\sigma \leq q$).
Step 2. Let $A_r = \{X \mid X \geq X^r\}$, for $r = 1, 2, \ldots, \sigma$, and compute the system reliability by using the sum of disjoint products (see Section 3.4 for more details).

3.5.2 BUDGET LIMIT

One may add some budget limit to the discussed problem in this section which makes the problem closer to the real-world cases. In such a case, it is enough to define a cost function for each SSV and then remove all the obtained (d,τ)-MP^{1}s which do not satisfy the budget constraint. It is noted that the definition of the cost function changes depending on which one of the transmission costs or the maintenance costs is considered.

3.5.3 A DESCRIPTIVE EXAMPLE

Consider the given network in Figure 3.1 with the given arc data, including the lead time and probability distribution for the arcs' capacities in Table 3.1, and assess its reliability for sending $d = 10$ units of the commodity within $\tau = 10$ units of time by using Algorithm 3.1.

Solution: According to Table 3.1, we have $M = (5, 3, 4, 1, 2, 4, 5, 1)$ and $L = (2, 2, 3, 1, 2, 2, 3, 1)$. There are nine MPs in the network: $P_1 = \{a_1, a_6\}$, $P_2 = \{a_1, a_4, a_7\}$, $P_3 = \{a_1, a_4, a_5, a_8\}$, $P_4 = \{a_2, a_4, a_6\}$, $P_5 = \{a_2, a_7\}$, $P_6 = \{a_2, a_5, a_8\}$, $P_7 = \{a_3, a_8\}$, $P_8 = \{a_3, a_5, a_7\}$, $P_9 = \{a_3, a_5, a_4, a_6\}$. From Equations (3.2) and (3.3), one can compute the capacity and lead time for each MP, then calculate the minimum integer

TABLE 3.1
The Arc Data for Figure 3.1

Arc	Lead Time	Capacity	Probability	Arc	Lead Time	Capacity	Probability
		5	0.7			2	0.85
		4	0.1	a_5	2	1	0.1
a_1	2	3	0.05			0	0.05
		2	0.05				
		1	0.05			4	0.7
		0	0.05			3	0.1
				a_6	2	2	0.1
		3	0.8			1	0.05
a_2	2	2	0.1			0	0.05
		1	0.05				
		0	0.05			5	0.7
						4	0.1
		4	0.7	a_7	3	3	0.05
		3	0.1			2	0.05
a_3	3	2	0.1			1	0.05
		1	0.05			0	0.05
		0	0.05				
				a_8	1	1	0.9
a_4	1	1	0.9			0	0.1
		0	0.1				

number c_j. Afterwards, if $c_j \leq CP_j$, the associated $(10,10)$-MP^1 with the MP can be computed by using Lemma 3.3. The final results of these calculations are given in Table 3.2. As it is seen in the table, we have only two $(10,10)$-MP's in this network which are determined in Step 1 of the algorithm. Now, in Step 2, letting $A_1 = \{X \geq X^1 = (2,0,0,0,0,2,0,0)\}$ and $A_2 = \{X \geq X^2 = (0,2,0,0,0,0,2,0)\}$, we have $R_{(d,\tau)} = \Pr(A_1) + \Pr(A_2) - \Pr(A_1 \cap A_2)$ which based on the given probability distributions for the arcs' capacities in Table 3.1 equals $R_{(d,\tau)} = 0.81 + 0.81 - 0.6561 = 0.9639$.

3.5.4 COMPLEXITY RESULTS

To compute the time complexity of Algorithm 3.1, we recall that m and q are, respectively, the number of arcs and MPs in the network. The required calculations corresponding to each MP in Step 1 are of order of $O(m)$, and hence the time complexity of this step is $O(mq)$. Assuming σ as the number of all the obtained (d,τ)-MP's in Step 1, the time complexity of calculating the reliability in Step 2 is $O(m\sigma^2)$ [39]. Therefore, as the steps are parallel, the time complexity of Algorithm 3.1 is $O(mq + m\sigma^2)$ which is less than or equal to $O(mq^2)$ as $\sigma \leq q$.

TABLE 3.2

The MPs and Corresponding (d,τ)-MP^1 s with $d = 10$ and $\tau = 10$ in Figure 3.1

j	P_j	CP_j	LP_j	c_j	$c_j \leq CP_j$?	Corresponding (d,τ)-MP^1
1	$\{a_1, a_6\}$	4	4	2	Yes	$X^1 = (2,0,0,0,0,2,0,0)$
2	$\{a_1, a_4, a_7\}$	1	6	3	No	–
3	$\{a_1, a_4, a_5, a_8\}$	1	6	3	No	–
4	$\{a_2, a_4, a_6\}$	1	5	2	No	–
5	$\{a_2, a_7\}$	3	5	2	Yes	$X^2 = (0,2,0,0,0,0,2,0)$
6	$\{a_2, a_5, a_8\}$	1	5	2	No	–
7	$\{a_3, a_8\}$	1	4	2	No	–
8	$\{a_3, a_5, a_7\}$	2	8	5	No	–
9	$\{a_3, a_5, a_4, a_6\}$	1	8	5	No	–

THEOREM 3.2

The time complexity of Algorithm 3.1 is at most $O(mq^2)$.

3.5.5 SOME OTHER ALGORITHMS

Here, we briefly described three literature algorithms that address the quickest path reliability problem with the need for any MPs in advance. Yeh et al. [11] proposed an algorithm that uses the kth-shortest path method to find all the (d,τ)-MP^1 s. The excellent point of the algorithm is no need for any MPs in advance; however, it uses several times the kth-shortest path technique, which makes the algorithm time-consuming. Yeh and Khadiri [22] employed the universal generating function approach to propose an algebraic algorithm to address the problem without the need for any MP in advance. The authors showed the superiority of the algorithm compared to the algorithm of Ref. [3]. Afterward, Yeh [21] employed the network characteristics and depth-first search method to propose a new algorithm to find all the (d,τ)-MP^1 s and showed that the algorithm is more efficient than the available algorithms in the literature.

3.6 TWO DISJOINT MINIMAL PATHS RELIABILITY PROBLEM

In the preceding section, the quickest path reliability problem was discussed in which the commodity is sent only through a single MP each time. However, the transmission time decreases if one simultaneously sends all the commodities through more than one MP. Hence, the problem has been extended to the cases with two DMPs [9, 19] and k DMPs, where $k \geq 2$ [10, 18]. The former is discussed here and the latter in the next section.

The primary extra assumption here is that the commodities are sent through two DMPs simultaneously. Accordingly, the adopted definition for the system reliability, denoted by $R_{(d,\tau)}^{(2)}$, is the probability of transmitting d units of a commodity from the source to the destination through two DMPs, P_i and P_j within τ units of time, where the system administrator gives the two DMPs in advance. The two DMPs reliability problem aims at computing such probability. Let us start our discussion with two definitions which are adapted from the definitions of similar concepts in the literature [9, 10].

DEFINITION 3.3

A system state vector (SSV) of X is called (d,τ)-MP^2 candidate, if d units of a commodity can be sent from the source to the destination through two DMPs, P_i and P_j under X within τ units of time.

DEFINITION 3.4

Assuming that Ψ_d is the set of all the (d,τ)-MP^2 candidates, then any $X \in \Psi_{d,\min}$ is called a (d,τ)-MP^2.

A notable difference between this problem and the one discussed in the preceding section is that now the given d units of a commodity can be transmitted through two DMPs, and accordingly one needs to divide d into two amounts of d_1 and d_2 which show the amount of commodity transmitted through each path. Hence, the vector $D = (d_1, d_2)$, where $d_1 + d_2 = d$, is called a *policy vector*. Without loss of generality and for more convenient, suppose that $P_1 = \{a_1, a_2, \ldots, a_{n1}\}$ and $P_2 = \{a_{n1+1}, a_{n1+2}, \ldots, a_{n1+n2}\}$ are the two specified MPs by the administrator. As these paths are disjoint, the transmission time for sending d units of a commodity from the source to the sink through these two paths under X and *policy vector* $D = (d_1, d_2)$ equals to

$$h^2(D, X) = \max\{h(P_1, d_1, X), h(P_2, d_2, X)\} \tag{3.8}$$

We recall that $h(P_j, d_j, X) = LP_j + \left\lceil \dfrac{d_j}{CP_j(X)} \right\rceil$, for $j = 1, 2$ (see Equation 3.4). Although the paths are given in advance, one needs to determine the *policy vector*, that is the values for d_1 and d_2. The following lemma provides an upper bound for each one.

LEMMA 3.4

The largest amount of commodity which can be transmitted through P_j within τ units of time is $\bar{d}_j = CP_j(M)(\tau - LP_j)$, for $j = 1, 2$.

Proof. We know that

$$h(P_j, \bar{d}_j, M) = LP_j + \left\lceil \frac{\bar{d}_j}{CP_j(M)} \right\rceil = LP_j + \left\lceil \frac{CP_j(M)(\tau - LP_j)}{CP_j(M)} \right\rceil = \tau.$$

It is noted that as $\tau - LP_j$ is an integer, $\lceil \tau - LP_j \rceil = \tau - LP_j$, and that we consider M as the current SSV as it contains the maximum capacity of each arc. Hence, $\bar{d}_j$ units of a commodity can be sent through P_j within time τ. Now, for $d = \bar{d}_j + \alpha > \bar{d}_j$, we have

$$h(P_j, d, M) = LP_j + \left\lceil \frac{d}{CP_j(M)} \right\rceil = LP_j + \left\lceil \frac{CP_j(M)(\tau - LP_j) + \alpha}{CP_j(M)} \right\rceil$$

$$= \tau + \left\lceil \frac{\alpha}{CP_j(M)} \right\rceil \geq \tau + 1 > \tau.$$

Therefore, the proof is complete.

Moreover, as $d_1 + d_2 = d$ and they are nonnegative values, we should have $d_1, d_2 \leq d$. Therefore, $\widehat{d}_j = \min\{d, \bar{d}_j\}$ is an upper bound for d_j, for $j = 1, 2$, and accordingly to find all the *policy vectors*, one needs to solve the following system.

$$\begin{cases} d_1 + d_2 = d \\ 0 \leq d_j \leq \widehat{d}_j, & \text{for } j = 1, 2. \end{cases} \tag{3.9}$$

PROPOSITION 3.3

The time complexity of solving the system (Equation 3.9) is $O(\delta)$, where $\delta = \max\{\widehat{d}_j \mid j = 1, 2\}$ is the maximum number of producible policy vectors.

The next issue is to determine the (d, τ)-MP^2 candidates and (d, τ)-MP^2 s corresponding to each *policy vector* (or producible from each *policy vector*). Reminding that the minimal candidates are the real (d, τ)-MP^2 s and following the same idea used in the preceding section (see Lemma 3.3), the following lemma provides a method to finds the smallest SSV under which d units of a commodity can be transmitted through paths P_1 and P_2 form the source to the destination within τ units of time and conforming the *policy vector* $D = (d_1, d_2)$.

LEMMA 3.5

Assume that the MPs, P_1 and P_2 and the policy vector $D = (d_1, d_2)$ are given, and

$$\tau > \max LP_1, LP_2. \ Let \ c_j = \left\lceil \frac{d_j}{\tau - LP_j} \right\rceil, for \ j = 1, 2. \ If \ c_j \le CP_j(M), for \ j = 1, 2, then$$

the vector $X = (x_1, x_2, \ldots, x_m)$ defined below is the smallest SSV associated with policy vector D under which d units of a commodity can be sent through P_1 and P_2 within τ units of time and conforming D.

$$For \ i = 1, 2, \ldots, m, x_i = \begin{cases} c_1 & if \ a_i \in P_1, c_2 \quad if \ a_i \in P_2, 0 \quad otherwise. \end{cases} \quad (3.10)$$

Proof. It can be seen that $c_j = \left\lceil \dfrac{d_j}{\tau - LP_j} \right\rceil$ is the smallest integer number such

that $LP_j + \left\lceil \dfrac{d_j}{c_j} \right\rceil \le \tau$. Hence, as $c_j \le CP_j(M)$, it is the minimum possible capacity

for P_j to transmit d_j units of a commodity through it within time τ. According to the definition of the capacity of an MP, it is clear that the above defined vector X is the smallest vector under which the capacity of P_j is equal to c_j, for $j = 1, 2$, and therefore, the proof is complete.

One sees that there is one producible (d, τ)-MP^2 candidate from each *policy vector*, and thus there are at most δ producible (d, τ)-MP^2 candidates. Although the obtained vectors from the above lemma are the smallest (d, τ)-MP^2 candidates corresponding to each *policy vector*, they are not necessarily the real (d, τ)-MP^2 s because they are not necessarily the minimal ones among all the producible candidates from all the *policy vectors*. Therefore, after determining all the candidates, one needs to remove the non-minimal ones. This removal is of order of $O(m\delta^2)$. As the only non-zero components in each produced candidate in Lemma 3.5 are c_1 and c_2, then one may compare the vectors $C = (c_1, c_2)$ to remove the non-minimal ones instead of comparing the corresponding candidates. This decreases the time complexity of removing the non-minimal ones from $O(m\delta^2)$ to $O(\delta^2)$. However, it is still not efficient as the number of (d, τ)-MP^2 candidates may grow exponentially. Next, an efficient approach is discussed, which has been originally proposed in Ref. [9].

3.6.1 AN EFFICIENT APPROACH TO DETERMINE NON-MINIMAL (d, τ)-MP^2 CANDIDATES

Assume that $P_1 = \{a_1, a_2, \ldots, a_{n1}\}$ and $P_2 = \{a_{n1+1}, a_{n1+2}, \ldots, a_{n1+n2}\}$ are the given DMPs, $D = (d_1, d_2)$ is the *policy vector* and $C = (c_1, c_2)$ is the vector defined in Lemma 3.5. Therefore, the obtained corresponding SSV is

$$X = (\underbrace{c_1, \ldots, c_1}_{n1 times}, \underbrace{c_2, \ldots, c_2}_{n2 times}, \underbrace{0, \ldots, 0}_{m-n1-n2 times})$$

If X is not a minimal (d,τ)-MP^2 candidate, then there exists another SSV, $Y < X$ which has been produced from another *policy vector* in similar way, and hence $Y = (c_1', \ldots, c_1', c_2', \ldots, c_2', 0, \ldots, 0)$. Without loss of generality, one can assume that $c_1' < c_1$ because $Y < X$. Therefore, there are numbers $d_1' < d_1$ and d_2' satisfying $c_1' = \left\lceil \dfrac{d_1'}{\tau - LP_1} \right\rceil$ and $c_2' = \left\lceil \dfrac{d_2'}{\tau - LP_2} \right\rceil$ which form the *policy vector* $D' = (d_1', d_2')$. It is clear that $d_2' > d_2$ because $d_1' < d_1$ and $d_1 + d_2 = d_1' + d_2'$, and thus c_2' cannot be less than c_2. Therefore, $c_2' = c_2$. As a result, the positive number $\alpha = d_1 - d_1'$ satisfies the following system.

$$\begin{cases} d_2' = d_2 + \alpha \\ \left\lceil \dfrac{d_1 - \alpha}{\tau - LP_1} \right\rceil < \left\lceil \dfrac{d_1}{\tau - LP_1} \right\rceil \\ \left\lceil \dfrac{d_2 + \alpha}{\tau - LP_2} \right\rceil = \left\lceil \dfrac{d_2}{\tau - LP_2} \right\rceil \end{cases} \tag{3.11}$$

Moreover, if $d_1 = 0$, then d_1' should be negative which is not acceptable and if $d_2 = 0$, then $c_2 = 0$ and consequently $c_2' = 0$ which is not possible. As a result, to find out if a vector $C = (c_1, c_2)$ is minimal, one needs to verify if it corresponds to a positive number α. The following theorem, proven in Ref. [9], is stated in this regard.

THEOREM 3.3

Given the MPs, P_1 and P_2, the demand value d and time limit τ, assume that $D = (d_1, d_2)$ is a policy vector such that $c_1 = \left\lceil \dfrac{d_1}{\tau - LP_1} \right\rceil$ and $c_2 = \left\lceil \dfrac{d_2}{\tau - LP_2} \right\rceil$ are both non-zero. In such a case,

1. we have $c_1' = \left\lceil \dfrac{d_1 - \alpha}{\tau - LP_1} \right\rceil < c_1$ and $c_2' = \left\lceil \dfrac{d_2 + \alpha}{\tau - LP_2} \right\rceil = c_2$, when
   a. $\alpha = d_1 - (c_1 - 1)(\tau - LP_1) > 0, \alpha + d_2 \leq c_2(\tau - LP_2)$.
2. *Or* similarly, we have $c_1' = \left\lceil \dfrac{d_1 + \alpha}{\tau - LP_1} \right\rceil = c_1$ and $c_2' = \left\lceil \dfrac{d_2 - \alpha}{\tau - LP_2} \right\rceil < c_2$, when
   b. $\alpha = d_2 - (c_2 - 1)(\tau - LP_2) > 0, d_1 + \alpha \leq c_1(\tau - LP_1)$.

Proof. Assuming $\alpha = d_1 - (c_1 - 1)(\tau - LP_1) > 0$ and $\alpha + d_2 \leq c_2(\tau - LP_2)$, we have $d_1 - \alpha = (c_1 - 1)(\tau - LP_1)$, and thus $c_1' = \left\lceil \dfrac{d_1 - \alpha}{\tau - LP_1} \right\rceil = c_1 - 1 < c_1$. Moreover, we have $d_2 + \alpha \leq c_2(\tau - LP_2)$, and hence $c_2' = \left\lceil \dfrac{d_2 + \alpha}{\tau - LP_2} \right\rceil = c_2$. Consequently, the first part is proven. One can prove the other part simply by exchanging the indices 1 and 2 in the proof of the first part.

To see if the vector $C = (c_1, c_2)$ is a minimal one, Theorem 3.3 verifies if there exists any vector $C' = (c_1', c_2') < (c_1, c_2) = C$. The vector C' is produced from the vector $D' = (d_1', d_2') = (d_1 - \alpha, d_2 + \alpha)$ or $(d_1 + \alpha, d_2 - \alpha)$. Hence, if D' does not satisfy the system (3.9), then it is not obtained, and accordingly C' is not generated. As a result, there is no need to remove the vector C. Hence, it should be first checked if the vector D' satisfies the system (3.9). Assuming that $D' = (d_1', d_2') = (d_1 - \alpha, d_2 + \alpha)$, since D is a *policy vector* which satisfies the system (3.9), it is vivid that $d_1' + d_2' = d_1 + d_2 = d$ and $d_1' = d_1 - \alpha \leq \widehat{d_1}$. Hence, one needs to only check for $d_2' = d_2 + \alpha \leq \widehat{d_2}$. Similarly, in the case of $D' = (d_1 + \alpha, d_2 - \alpha)$, one needs to check for $d_1' = d_1 + \alpha \leq \widehat{d_1}$.

Consequently, one concludes the following approach to determine the non-minimal C vectors. If $C = (c_1, c_2)$ is calculated as described in Lemma 3.5 from $D = (d_1, d_2)$, then C is non-minimal when both d_1 and d_2 are positive and one of the following holds:

1. $\alpha_1 = d_1 - (c_1 - 1)(\tau - LP_1) > 0$, $\alpha_1 + d_2 \leq c_2(\tau - LP_2)$ and $\alpha_1 + d_2 \leq \widehat{d_2}$.
2. $\alpha_2 = d_2 - (c_2 - 1)(\tau - LP_2) > 0$, $d_1 + \alpha_2 \leq c_1(\tau - LP_1)$ and $d_1 + \alpha_2 \leq \widehat{d_1}$.

The time complexity for computing α and checking the above cases is $O(1)$. Hence, with δ being the number of obtained *policy vectors*, the time complexity to find and remove the non-minimal vectors by using the above-proposed technique is $O(\delta)$ which is less than $O(\delta^2)$ that is the time complexity of the comparative approach used in Ref. [19]. Therefore, the proposed method is more efficient compared to the comparative technique.

LEMMA 3.6

The abovementioned technique needs $O(\delta)$ time to check all the vectors $C = (c_1, c_2)$ for being minimal ones, where $c_j = \left\lceil \dfrac{d_j}{\tau - LP_j} \right\rceil$, for $j = 1, 2,$ and δ is the number of all the obtained policy vectors.

We note that some duplicate C vectors are not removed in the above-stated technique. Thus, comparing the final minimal C vectors to remove the duplicates is required. To do so, instead of comparing the vectors, one can use a more efficient approach proposed by Forghani-elahabad and Mahdavi-Amiri [8]. For more convenience, it is explained here how one can use this technique in our case.

Assume that $C = (c_1, c_2)$ is obtained by using Lemma 3.5, k_j is the number of digits in c_j, for $j = 1, 2$ and k is the number of digits in d. It is clear that $k_j \leq k$ as $d_j \leq d$, for $j = 1, 2$. Next, an k-digit number, that is n_j, is constructed for each component in the vector of C. If $k = k_j$, then let $n_j = c_j$, and if $k_j < k$, then let $n_j = \underbrace{00 \cdots 0}_{k-k_j \text{ times}} c_j$. This way, one can construct a $2k$-digit number for C as $N_C = n_1 n_2$ associated with the vector C. As there is a bijective relationship between the vectors and the numbers, one can find the duplicates by comparing the numbers instead of the vectors [8]. The time complexity of this approach to find the duplicates among δ vectors is $O(\delta \log(\delta))$ [8].

3.6.2 AN ALGORITHM TO ADDRESS THE PROBLEM

By using the stated results in this section, the following algorithm based on the one proposed in Ref. [9] is stated to address the two DMPs reliability problem.

Algorithm 3.2 (network reliability evaluation based on (d,τ)-MP^2 s)
Input: A multistate flow network with two DMPs, P_1 and P_2, a time limit τ and a demand level d.
Output: The system reliability, that is $R_{(d,\tau)}^{(2)}$.

Step 1. Compute $CP_j(M) = \min\{M_i \mid a_i \in P_j\}$, $LP_j = \sum_{a_i \in P_j} l_i$ and $\widehat{d}_j = \min\{d,$

$CP_j(M)(\tau - LP_j)\}$, for $j = 1, 2$. Let $S_1 = S_2 = \phi$ and $\sigma = 0$ (the number of (d,τ)-MP^2 s).

Step 2. If $\widehat{d}_1 + \widehat{d}_2 < d$ then go to Step 7, else let $r = 0$, $r_{\max} = \min\left\{\widehat{d}_1, \widehat{d}_1 + \widehat{d}_2 - d\right\}$ and $(fd_1, fd_2) = (\widehat{d}_1, d - \widehat{d}_1)$.

Step 3. If $r > r_{\max}$ then go to Step 7 else let $D = (d_1, d_2) = (fd_1 - r, fd_2 + r)$,

$$c_1 = \left\lceil \frac{d_1}{\tau - LP_1} \right\rceil, c_2 = \left\lceil \frac{d_2}{\tau - LP_2} \right\rceil, C = (c_1, c_2) \text{ and } r = r + 1.$$

Step 4. Determine the corresponding number to C, that is N_C. If it is a duplicate number, then remove the number together with the vector C and go to Step 3 else add N_C into S_1.

Step 5. (To Check C for being a minimal vector)

5.1. If $c_1 = 0$ or $c_2 = 0$ then go to Step 6.

5.2. If $\alpha_1 = d_1 - (c_1 - 1)(\tau - LP_1) > 0$, $\alpha_1 + d_2 \le c_2(\tau - LP_2)$, and $\alpha_1 + d_2 \le \widehat{d}_2$, then C is non-minimal, remove it and go to Step 3.

5.3. If $\alpha_2 = d_2 - (c_2 - 1)(\tau - LP_2) > 0$, $d_1 + \alpha_2 \le c_1(\tau - LP_1)$ and $d_1 + \alpha_2 \le \widehat{d}_1$, then C is non-minimal, remove it and go to Step 3.

Step 6. Let $\sigma = \sigma + 1$, and then according to Lemma 3.5, set the vector $X^\sigma = (x_1, \ldots, x_m)$ as follows:

$$\text{For } i = 1, 2, \ldots m, x_i = \begin{cases} c_1, & \text{if } a_i \in P_1, \\ c_2, & \text{if } a_i \in P_2, \\ 0, & \text{otherwise.} \end{cases}$$

Add X^σ into S_2 and go to Step 3.

Step 7. If $\sigma = 0$, then halt, it is not possible to transmit d units of a commodity within τ units of time through the given MPs. Otherwise, $S_2 = \left\{X^1, X^2, \ldots, X^\sigma\right\}$ is the set of all the (d,τ)-MP^2 s. Let $A_r = \{X \mid X \ge X^r\}$, for $r = 1, 2, \ldots, \sigma$, and compute the system reliability by using the sum of disjoint products (see Section 3.4 for more details).

THEOREM 3.4

Algorithm 3.2 *determines all the* (d,τ)-MP^2s *and evaluates the network reliability, that is* $R_{(d,\tau)}$, *correctly.*

Proof. The algorithm has two main stages: finding all the (d,τ)-MP^2s and evaluating the reliability using the obtained (d,τ)-MP^2 s. As it uses the known techniques from the literature in the second stage, one needs to prove only the correctness of the first stage. To this aim, without loss of generality, let $P_1 = \{a_1,...,a_{n1}\}$ and $P_2 = \{a_{n1+1},...,a_{n1+n2}\}$ be the given MPs. It is shown that (1) every obtained vector by the algorithm is a (d,τ)-MP^2, and (2) the algorithm determines all the (d,τ)-MP^2 s.

(1): Let $X^* \in S_2$ is a vector obtained by Algorithm 3.2 with corresponding MPs capacity vector $C^* = (c_1,c_2)$ (see Step 5) and *policy vector* $D^* = (d_1,d_2)$ (see Step 2). Therefore, according to Step 5, we have $X^* = (\underbrace{c_1,...,c_1}_{n1\,times},\underbrace{c_2,...,c_2}_{n2\,times},\underbrace{0,...,0}_{m-n1-n2\,times})$. To show that X^* is a (d,τ)-MP^2, on the contrary, assume that it is not. Hence, there exists a (d,τ)-MP^2, say $Y = (y_1,...,y_m) < X^* = (x_1,...,x_m)$. Hence, there is a $1 \le i \le m$ so that $y_j < x_j$. Without loss of generality, let $1 \le j \le n1$, that is $a_j \in P_1$. From Step 5, we know that $x_i = c_1$, and Lemma 3.5 says that c_1 is the smallest possible capacity for P_1 to make it possible to send d_1 units of a commodity through P_1 within τ units of time. As $y_i < x_i$ and $a_i \in P_1$, one has $CP_1(Y) = \min\{y_r \mid a_r \in P_1\} \le y_i < x_i = c_1$, and consequently $h(P_1,d_1,Y) > \tau$ which contradicts with Y being a (d,τ)-MP^2. Hence, the first part of proof is done.

(2): Suppose that $X' = (x_1,...,x_m)$ is a (d,τ)-MP^2 correspond to *policy vector* $D = (d_1,d_2)$. We prove that X' can be obtained by Algorithm 3.2. If there is a $n1+n2 < j \le m$ such that $x_j > 0$, then let $Y = (x_1,...,x_{j-1},x_j-1,x_{j+1},...,x_m) < X'$. It is easy to see that $CP_1(Y) = CP_1(X')$ and $CP_2(Y) = CP_2(X')$ as a_j does not belong to P_1 or P_2. Hence, $h^2(D,Y) = h^2(D,X') \le \tau$ which makes Y a (d,τ)-MP^2 candidates, contradicting with X' being a (d,τ)-MP^2. Therefore, for each $n1+n2 < j \le m$, we have $x_j = 0$. Now, let $c'_1 = CP_1(X')$, $c'_2 = CP_2(X')$, $C' = (c'_1,c'_2)$ and $X^* = (\underbrace{c'_1,...,c'_1}_{n1\,times},\underbrace{c'_2,...,c'_2}_{n2\,times},\underbrace{0,...,0}_{m-n1-n2\,times})$. According to the definition of the capacity of an MP and the values of c'_1 and c'_2, it is easy to see that $X^* \le X'$. If $X^* < X'$, as $CP_1(X^*) = CP_1(X')$ and $CP_2(X^*) = CP_2(X')$, one can conclude that X^* is a (d,τ)-MP^2 candidate, contradicting with X' being a (d,τ)-MP^2. So, $X^* = X'$. It shows that X' can be driven from C' in Step 5 of Algorithm 3.2. Moreover, as C' is a vector which denotes the capacities of P_1 and P_2, Lemma 3.5 shows that there should be a *policy vector* D' corresponding to C'. On the other hand, we know that all the possible *policy vectors* are obtained in Step 2 of Algorithm 3.2. Thus, the vectors C' and X' are generated by Algorithm 3.2, and accordingly the proof is complete.

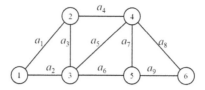

FIGURE 3.2 An adjusted benchmark network example taken from Refs. [23, 40].

Next, let us illustrate the algorithm through a benchmark network example.

3.6.3 A DESCRIPTIVE EXAMPLE

Consider the network given in Figure 3.2 with the arcs' data provided in Table 3.3. Assume that the DMPs, $P_1 = \{a_1, a_4, a_8\}$ and $P_2 = \{a_2, a_6, a_9\}$ are determined and the administrator wants to know the system reliability for transmitting $d = 6$ units of a commodity from node 1 to node 6 through these paths within $\tau = 10$ units of time. We solve the problem by using Algorithm 3.2 as follows.

TABLE 3.3
The Arc Data for Figure 3.2

Arc	Lead Time	Capacity	Probability	Arc	Lead Time	Capacity	Probability
		4	0.75			2	0.85
		3	0.1	a_6	2	1	0.1
a_1	2	2	0.05			0	0.05
		1	0.05				
		0	0.05			2	0.9
				a_7	1	1	0.05
		3	0.8			0	0.05
a_2	3	2	0.1				
		1	0.05			4	0.7
		0	0.05			3	0.1
				a_8	3	2	0.1
a_3	2	1	0.9			1	0.05
		0	0.1			0	0.05
		2	0.9			3	0.8
a_4	4	1	0.05	a_9	2	2	0.1
		0	0.05			1	0.05
						0	0.05
		2	0.9				
a_5	3	1	0.05				
		0	0.05				

Step 1. We have $CP_1 = 2$, $CP_2 = 2$, $LP_1 = 9$, $LP_2 = 7$, $\widehat{d}_1 = 2$, $\widehat{d}_2 = 6$, $S_1 = S_2 = \phi$
and $\sigma = 0$.

Step 2. $\widehat{d}_1 + \widehat{d}_2 = 8 \nless d = 6$. Let $r = 0$, $r_{max} = 2$ and $(fd_1, fd_2) = (2,4)$.

Step 3. $r = 0 \ngtr 2 = r_{max}$. Let $D = (2,4)$, $c_1 = 2$, $c_2 = 2$ and $r = 1$.

Step 4. We have $N_C = 22$ which is not duplicate, and so $S_1 = \{22\}$.

Step 5. (Checking C for being a minimal vector)

  5.1 $c_1 \neq 0, c_2 \neq 0$.

  5.2 $\alpha_1 = 1 > 0$, $\alpha_1 + d_2 = 5 \leq 6 = c_2(\tau - LP_2)$ and $\alpha_1 + d_2 = 5 \leq 6 = \widehat{d}_2$. Hence,
    C is not a minimal vector and we go to Step 3.

Step 3. $r = 1 \ngtr 2 = r_{max}$. Let $D = (1,5)$, $c_1 = 1$, $c_2 = 2$ and $r = 2$.

Step 4. We have $N_C = 12$ which is not duplicate, and so $S_1 = \{22,12\}$.

Step 5. (Checking C for being a minimal vector)

  5.1 $c_1 \neq 0, c_2 \neq 0$.

  5.2 $\alpha_1 = 1 > 0$, $\alpha_1 + d_2 = 6 \leq 6 = c_2(\tau - LP_2)$ and $\alpha_1 + d_2 = 6 \leq 6 = \widehat{d}_2$. Hence,
    C is not a minimal vector and we go to Step 3.

Step 3. $r = 2 \ngtr 2 = r_{max}$. Let $D = (0,6)$, $c_1 = 0$, $c_2 = 2$ and $r = 3$.

Step 4. We have $N_C = 02 = 2$ which is not duplicate, and so $S_1 = \{22,12,2\}$.

Step 5. (Checking C for being a minimal vector)

  5.1 $c_1 = 0$, and so we go to Step 6.

Step 6. We have $\sigma = 1$, $X^1 = (0,2,0,0,0,2,0,0,2)$ and $S_2 = \{X^1\}$ and go to
Step 3.

Step 3. $r = 3 > 2 = r_{max}$, and so we go to Step 7.

Step 7. $\sigma \neq 0$, and so $S_2 = \{X^1\}$ is the set of all the $(6,10)$-MP^2 in this network.
According to the discussion in Section 3.4 and using the probability distributions given in Table 3.3, the system reliability equals to $R_{6,10} = 0.6885$
which is not so good.

3.6.4 COMPLEXITY RESULTS

The time complexity of Step 1 to calculate the required data is $O(m)$. Steps 2 and 3
are clearly of order of $O(1)$. Considering δ as the number of generated *policy vectors*,
the time complexity of Steps 4, 5, and 6 are, respectively, $O(\delta \log \delta), O(\delta)$, and $O(m\delta)$.
We recall that $\delta = r_{max} + 1$ in this algorithm. Therefore, the time complexity of
Steps 3–5 is $O((m + \log \delta)\delta)$. As σ is the number of all the obtained (d,τ)-MP^2 s, the
time complexity of calculating the reliability in Step 7 is $O(m\sigma^2)$ [39]. Since Steps 0–2,
Steps 3–6, and Step 7 are executed in parallel with each other, the time complexity
of Algorithm 3.2 is $O((m + \log \delta)\delta + m\sigma^2)$.

THEOREM 3.5

The time complexity of Algorithm 3.2 is $O((m + \log \delta)\delta + m\sigma^2)$.

3.6.5 MORE DISCUSSION

3.6.5.1 Another Algorithm

We note that Algorithm 3.2 removes the duplicate vectors in Step 4 and recognizes the non-minimal (d,τ)-MP^2 s in Step 5. One can remove these two steps from the algorithm and instead add a new step after Step 6 in the algorithm for comparing all the (d,τ)-MP^2 candidates to clear the non-minimal ones together with duplicates. This way, the algorithm will be similar to the proposed algorithm in Ref. [19] which is less efficient than Algorithm 3.2 [9].

3.6.5.2 Budget Limit

Similar to the quickest path reliability problem, one can add some budget limits to this problem. In such a case, one may use Algorithm 3.2 and add a new step to check each obtained (d,τ)-MP^2 for budget limit. On the other hand, one may use such a limit to decrease the number of generated *policy vectors* and candidates as it has been done in Ref. [9].

3.7 SEVERAL DISJOINT MINIMAL PATHS RELIABILITY PROBLEM

The preceding section studied the extension of the quickest path reliability problem to the case with two DMPs. Here, the aim is to generalize it to the case with k DMPs, where $k \geq 2$. Therefore, most of the definitions and results here are similar to the ones from the preceding section. However, this problem is discussed and addressed in this chapter because it has been studied in the literature individually [10, 18]. Moreover, we know that increasing the number of used DMPs can decrease the transmission time. We note that the adopted definition for the system reliability, denoted by $R_{(d,\tau)}^{(k)}$, is the probability of sending d units of a commodity from the source to the sink through $k \geq 2$ DMPs within time τ. Since the focus is on reliability evaluation, it is assumed that the DMPs, $P_1,\ldots,P_k$ are determined and given in advance. Let us start with two definitions which are adapted from the definitions of similar concepts in the literature [9, 10].

DEFINITION 3.5

A system state vector (SSV) of X is called (d,τ)-MP^k candidate, if d units of a commodity can be transmitted from the source to the destination through the DMPs, $P_1,\ldots,P_k$ within τ units of time under X.

DEFINITION 3.6

Assuming that Ψ_d is the set of all the (d,τ)-MP^k candidates, then any SSV, $X \in \Psi_{d,\min}$ is called a (d,τ)-MP^k.

We remind that the set $X \in \Psi_{d,\min}$ is the set of all the minimal vectors in the set Ψ_d, and that $X \in \Psi_d$ is a minimal vector when there is no any vector $Y \in \Psi_d$ such that

$Y < X$. Similar to the case with two MPs, here one needs to divide the demand value d into k sub-demand values $d_1,\ldots,d_k$, and accordingly the vector $D = (d_1,\ldots,d_k)$ is called a *policy vector*. As the commodities are transmitting through the paths simultaneously and the paths are disjoint, it is clear that the transmission time for sending $d = d_1 + \cdots + d_k$ units of a commodity through these paths under SSV of X is equal to

$$h^k(D,X) = \max\{h(P_1,d_1,X),\ldots,h(P_k,d_k,X)\}, \tag{3.12}$$

where $h(P_j,d_j,X) = LP_j + \left\lceil \dfrac{d_j}{CP_j(X)} \right\rceil$ is the transmission time for sending d_j units of a commodity through P_j, for $j = 1,\ldots,k$. The following lemma is similar to Lemma 3.4, and so is stated without proof.

LEMMA 3.7

The largest amount of commodity which can be transmitted through P_j within τ units of time is $\bar{d}_j = CP_j(M)(\tau - LP_j)$, for $j = 1,\ldots,k$.
 Similarly, letting $\hat{d}_j = \min\{d,\bar{d}_j\}$, for $j = 1,\ldots,k$, the corresponding system to find all the *policy vectors* is the following.

$$\begin{cases} d_1 + \cdots + d_k = d \\ 0 \le d_j \le \hat{d}_j, \quad \text{for } j = 1,\ldots,k. \end{cases} \tag{3.13}$$

The system above is a Diophantine system for which an efficient approach has been provided in Ref. [14]. Letting δ be the number of obtained *policy vectors*, the time complexity of solving the system (3.13) is $O(k\delta)$ (see Section 3.1 in Ref. [14]). Having a *policy vector*, the next step is to find its corresponding smallest SSV. The following lemma provides an approach to calculate such a vector. As it is similar to the case with two disjoint paths (see Lemma 3.5), the proof is left to the reader.

LEMMA 3.8

Let $P_1,\ldots,P_k$ be the DMPs and $D = (d_1,\ldots,d_k)$ be a policy vector for which

$\tau > \max\{LP_1,\ldots,LP_k\}$. *Then, set* $c_j = \left\lceil \dfrac{d_j}{\tau - LP_j} \right\rceil$, *for* $j = 1,\ldots,k$. *If* $c_j \le CP_j(M)$,

for $j = 1,\ldots,k$, then the vector $X = (x_1,x_2,\ldots,x_m)$ defined below is the smallest SSV associated with D under which d units of a commodity can be sent through the paths $P_1,\ldots,P_k$ (d_j units through P_j) within time τ.

For $i = 1,2,\ldots,m, x_i = \begin{cases} c_1 & \text{if } a_i \in P_1, \vdots & \vdots c_k & \text{if } a_i \in P_k, 0 & \text{otherwise.} \end{cases}$ (3.14)

It is recalled that the defined vector in the lemma above is the smallest producible (d,τ)-MP^k candidate corresponding to *policy vector D*; however, it is not necessarily a minimal vector among all the candidates generated from all the *policy vectors*. Therefore, one needs to remove all the non-minimal candidates to have at the (d,τ)-MP^ks. An obvious but not so efficient approach is to compare all the candidates. As the number of DMPs, that is k, in a network is really less than the number of arcs, that is m, a better approach is to compare the C vectors instead of candidates because they are k-tuple whereas the candidates are m-tuple vectors. Here, generalizing Theorem 3.3, we provide an efficient technique to recognize the non-minimal C vectors. A similar result is proven in Ref. [10].

THEOREM 3.6

Let $D = (d_1, d_2, \ldots, d_k)$ *be a policy vector, that is* $d_1 + d_2 + \cdots + d_k = d$, *with corresponding vector of* $C = (c_1, \ldots, c_k)$ *where* $c_r = \left\lceil \dfrac{d_r}{\tau - LP_r} \right\rceil$, *for* $r = 1, 2, \ldots, k$.

Then, C is a non-minimal vector if and only if for at least one $1 \le j \le k$, $s_j = d_j - (c_j - 1)(\tau - LP_j)$ *is positive and the following system has integer solutions* γ_i, *for* $i = 1, \ldots, k$ *and* $i \ne j$:

$$
\begin{cases}
\text{(i)} \displaystyle\sum_{1 \le i \le k,\, i \ne j} \gamma_i = s_j, \\[4mm]
\text{(ii)}\, 0 \le \gamma_i \le u_i, \qquad i = 1, 2, \ldots, k, \text{and } i \ne j,
\end{cases}
\tag{3.15}
$$

where $u_i = \min\left\{ \widehat{d_i}, c_i(\tau - LP_i) \right\} - d_i$, *for* $i = 1, 2, \ldots, k$ *and* $i \ne j$.

Proof. $\Rightarrow$: Let $C = (c_1, \ldots, c_k)$ be a non-minimal vector corresponding to $D = (d_1, d_2, \ldots, d_k)$. Thus, there exists a vector $C' = (c_1', \ldots, c_k') < C = (c_1, \ldots, c_k)$. Therefore, $c_i' \le c_i$, for $i = 1, \ldots, k$, and there exists at least one $j = 1, \ldots, k$ such that $c_j' < c_j$. Let D' be the corresponding *policy vector* to C'. Hence, it is easy to see that $d_j' < d_j$. Let $\gamma = d_j - d_j'$ which is a positive number. Since $d_1 + \cdots + d_k = d_1' + \cdots + d_k'$, there exist $u \ge 1$ components of the vectors D and D', say d_{i_r} and d_{i_r}', for $r = 1, \ldots, u$, satisfying $d_{i_r}' \ge d_{i_r}$, for $r = 1, \ldots, u$, and $\displaystyle\sum_{r=1}^{u} d_{i_r}' = \sum_{r=1}^{u} d_{i_r} + \gamma$. It is noted that $i_r \ne j$, for $r = 1, \ldots, u$. Also, since $C' < C$, we have $c_{i_r}' \le c_{i_r}$, for $r = 1, \ldots, u$. Therefore, $c_{i_r}' = c_{i_r}$, for $r = 1, \ldots, u$ (it is reminded that $c_{i_r} = \left\lceil \dfrac{d_{i_r}}{T - LP_{i_r}} \right\rceil$ and $c_{i_r}' = \left\lceil \dfrac{d_{i_r}'}{T - LP_{i_r}} \right\rceil$, for $r = 1, \ldots, u$). Consequently, one can increase the components d_{i_r} to d_{i_r}', for $r = 1, \ldots, u$, whereas the corresponding components c_{i_r} do not change. Thus, one can start from the vector D, enhance the components d_{i_r}, one by one, limited to d_{i_r}', for $r = 1, \ldots, u$, and decrease d_j by one unit concurrently with each unit of increment in d_{i_*}. This process is continued till c_j is lessened by one unit. As a result of this procedure, a *policy vector* of D'' will be at hand with the corresponding vector of $C'' = (c_1'',\dots,c_k'')$ which satisfies $c_j'' = c_j - 1$ and $c_i'' = c_i$, for $i = 1,\dots,k$ and $i \neq j$. Now, one observes that $s_j = d_j - d_j'' = d_j - (c_j - 1)(T - LP_j)$ is positive and $\gamma_i = d_i'' - d_i$, for $i = 1,2,\dots,q$, and $i \neq j$ are the integer solutions of the system (3.15). This completes the first part.

$\Leftarrow$: Let $D = (d_1,\dots,d_k)$ be a *policy vector* associated with the vector $C = (c_1,\dots,c_k)$, and suppose that for one $1 \leq j \leq q$, $s_j = d_j - (c_j - 1)(T - LP_j)$ is positive and that γ_i, for $i = 1,2,\dots,q$, and $i \neq j$ are the integer solutions of the system (3.15). Let $d_j' = d_j - s_j$, $d_i' = d_i + \gamma_i$, for $i = 1,2,\dots,q$, and $i \neq j$, and $D' = (d_1',\dots,d_k')$. From the system (3.15), it is resulted that D' is a *policy vector*. Moreover, from the definitions of s_j and u_i s, one can conclude that $C' = (c_1,\dots,c_{j-1},c_j-1,c_{j+1},\dots,c_k)$ is the corresponding vector to D', and hence C is a non-minimal vector.

We note that it is not required to solve the system (3.15), and in fact one needs only to check if the system has any solutions or not. One easily observes that the system has solutions if and only if $\displaystyle\sum_{i=1,i\neq j}^{k} u_i \geq s_j$. Hence, this condition can be checked instead of searching for any solutions for the system. Moreover, to avoid repeating the calculation of this summation for each $j = 1,\dots,k$, one can calculate once $\bar{u} = \displaystyle\sum_{i=1}^{k} u_i$, and then for each j, check the condition $\bar{u} - u_j \geq s_j$.

A weak point of applying the theorem above is that it does not find the duplicate minimal vectors, and hence one needs to compare the minimal vectors, namely (d,τ)-MP^ks, to remove the duplicates. However, as the number of (d,τ)-MP^ks is far less than the number of candidates, applying the theorem is still more efficient than the comparative approach. Moreover, to remove the duplicates, one can use the proposed method in Ref. [8], which was described in the preceding section, to compare the numbers instead of the vectors.

3.7.1   AN ALGORITHM TO ADDRESS THE PROBLEM

Now that we know how to generate the *policy vectors*, find the corresponding (d,τ)-MP^k candidates, remove the non-minimal ones, and determine the (d,τ)-MP^ks, we are at the point to state the algorithm. Note that the following algorithm is based on the proposed algorithm in Ref. [10] for the same problem.

Algorithm 3.3 (network reliability evaluation based on (d,τ)-MP^ks)
Input: A multistate flow network with k DMPs, demand level d and time
   limit τ.
Output: The network reliability, that is $R_{(d,\tau)}^{(k)}$.

Step 1. Calculate $LP_j = \sum_{a_i \in P_j} l_i$, $CP_j(M) = \min\{M_i \mid a_i \in P_j\}$ and $\widehat{d}_j = \min\{d,$

$CP_j(M)(\tau - LP_j)\}$, for $j = 1, 2, \ldots, k$. Let $S_1 = S_2 = S_3 = \phi$ and $\sigma = 0$ (the number of final (d, τ)-MP^ks).

Step 2. Find the next *policy vector* $D = (d_1, d_2, \ldots, d_k)$ by solving the following system:

$$\begin{cases} (1)\, d_1 + d_2 + \cdots + d_k = d, \\ (2)\, 0 \le d_j \le \widehat{d}_j, \qquad\qquad j = 1, 2, \ldots, k, \\ (3)\, d_j \text{ sare integers.} \end{cases}$$

If there is no solution, then go to Step 8.

Step 3. Let $c_j = \left[\dfrac{d_j}{\tau - LP_j} \right]$, for $j = 1, \ldots, k$, and $C = (c_1, \ldots, c_k)$.

Step 4. Generate the number N_C associated with the vector C. If it is a duplicate number, then go to Step 2, else add N_C to S_1.

Step 5. (To Check C for being a minimal vector)

5.1. Calculate $u_i = \min\{\widehat{d}_i, c_i(\tau - LP_i)\} - d_i$, for $i = 1, 2, \ldots, k$, and $\overline{u} = \sum_{i=1}^{k} u_i$.
If $\overline{u} = 0$, then go to Step 6, else let $j = 1$.

5.2. Calculate $s_j = d_j - (c_j - 1)(\tau - LP_j)$. If $s_j > 0$ and $\overline{u} - u_j \ge s_j$, then go to Step 2 ($C$ is a non-minimal vector).

5.3. If $j < k$, then let $j = j + 1$ and go to Step 5.2.

Step 6. (To generate the state vector X corresponding to C) For $i = 1, 2, \ldots m$, set $x_i = \lambda_i \ge c_j$, if $a_i \in P_j$, for any $j = 1, \ldots, k$, where λ_i is the smallest possible capacity for a_i greater than or equal to c_j. Then, let $X = (x_1, x_2, \ldots, x_m)$.

Step 7. Compute the number N_X associated with X. If it is not a duplicate number, then let $\sigma = \sigma + 1$, $X^\sigma = X$, add N_X to S_2 and X^σ to S_3. Go to Step 2.

Step 8. If $\sigma = 0$, then halt, it is not possible to send d units of a commodity within τ units of time through the given DMPs. Otherwise, $S_3 = \{X^1, X^2, \ldots, X^\sigma\}$ is the set of all the (d, τ)-MP^ks. Let $A_r = \{X \mid X \ge X^r\}$, for $r = 1, 2, \ldots, \sigma$, and compute the system reliability by using the sum of disjoint products (see Section 3.4 for more details).

We note that as it is possible to generate duplicate (d, τ)-MP^ks from different C vectors, one needs to check (d, τ)-MP^ks for duplicates even when all the duplicate C vectors have already been removed. Hence, Step 7 is required in the algorithm. As Algorithm 3.3 is based on Definitions 3.5 and 3.6, Lemmas 3.7 and 3.8, Theorem 3.6 and uses the sum of disjoint products technique, the correctness of the algorithm is clear and the following result is at hand.

THEOREM 3.7

Algorithm 3.3 finds all the (d, τ)-MP^ks and evaluates the network reliability correctly.

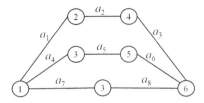

FIGURE 3.3   A network example with three DMPs.

A notable point about the cases with two or more DMPs is that none of the arcs which do not belong to the given DMPs affect the system reliability because no commodity is transmitted through these arcs. Accordingly, one can remove these arcs from the network. Therefore, in the following example, we provide a network with only the arcs which belong to the given paths to illustrate Algorithm 3.3.

3.7.2   A Descriptive Example

Consider the network given in Figure 3.3 with three DMPs, $P_1 = \{a_1, a_2, a_3\}$, $P_2 = \{a_4, a_5, a_6\}$ and $P_3 = \{a_7, a_8\}$. The arcs' data of this network is provided in Table 3.4. The administrator wants to know the system reliability for transmitting $d = 14$ units of a commodity from node 1 to node 8 through these paths within $\tau = 9$ units of time. We solve the problem by using Algorithm 3.3 as follows.

> **Step 1.** We have $LP_1 = 7, LP_2 = 8, LP_3 = 8, CP_1(M) = 4, CP_2(M) = 4, CP_3(M) = 5$, $\hat{d}_1 = 8, \hat{d}_2 = 4$, and $\hat{d}_3 = 5$. Let $S_1 = S_2 = S_3 = \phi$ and $\sigma = 0$.
> **Step 2.** Using the proposed algorithm in Ref. [14], the first *policy vector* $D = (5, 4, 5)$ is obtained.
> **Step 3.** We have $C = (3, 4, 5)$.
> **Step 4.** $N_C = 345$ is not duplicate, and so $S_1 = \{345\}$.
> **Step 5.** Checking C for being minimal:
>> **Step 5.1.** We have $u_1 = 1, u_2 = 0, u_3 = 0$, and $\bar{u} = 1 \neq 0$, and so let $j = 1$.
>> **Step 5.2.** We have $s_1 = d_1 - (c_1 - 1)(\tau - LP_1) = 1$ which is positive but $\bar{u} - u_1 = 0 \not\geq 1 = s_1$.
>> **Step 5.3.** As $j = 1 < 3 = k$, let $j = 2$ and go to Step 5.2.
>> **Step 5.2.** We have $s_2 = d_2 - (c_2 - 1)(\tau - LP_2) = 1$ which is positive and $\bar{u} - u_2 = 1 \geq 1 = s_1$, and so we go to Step 2 because C is not a minimal vector.
> **Step 2.** The next obtained solution is $D = (6, 3, 5)$.
> **Step 3.** We have $C = (3, 3, 5)$.
> **Step 4.** $N_C = 335$ is not duplicate, and so $S_1 = \{345, 335\}$.
> **Step 5.** Checking C for being minimal:
>> **Step 5.1.** We have $u_1 = u_2 = u_3 = 0$, and so $\bar{u} = 0$. So, we go to Step 6.
> **Step 6.** The corresponding SSV to $C = (3, 3, 5)$ is the vector $X = (3, 3, 4, 3, 4, 4, 5, 6)$.
> **Step 7.** $N_X = 334344556$ is not duplicate, and so we have $\sigma = 1, X^1 = (3, 3, 4, 3, 4, 4, 5, 6)$, $S_2 = \{334344556\}$, and $S_3 = \{X^1\}$. Then, go to Step 2.

Step 8. The final set of all the $(14,9)$-MP^3 is equal to $S_3 = \{$ $(3, 3, 2, 3, 4, 4, 5, 6)$, $(3, 3, 4, 5, 4, 4, 5, 4)$, $(5, 5, 4, 5, 4, 4, 2, 2)$, $(5, 5, 4, 3, 4, 4, 3, 4)$, $(5, 5, 4, 3, 2, 2, 5, 6)$, $(5, 5, 4, 1, 1, 1, 5, 6)$ $\}$. Accordingly, using the sum of disjoint products and given arcs data in Table 3.4, the network reliability is equal to 0.83442.

3.7.3 COMPLEXITY RESULTS

The calculations in Step 1 are of order of $O(mk)$. Solving the system in Step 2 and finding all the *policy vectors* by using the proposed algorithm in Ref. [14] is of order of $O(k\delta)$ where δ is the number of *policy vectors* (see Section 3.1 in Ref. [14]). The time complexity of Step 3 for each *policy vector* is $O(k)$, and so for all the *policy*

TABLE 3.4

The Arc Data for Figure 3.3

Arc	Lead Time	Capacity	Probability	Arc	Lead Time	Capacity	Probability
		5	0.95			6	0.95
		4	0			5	0
a_1	2	3	0,03	a_5	3	4	0.02
		2	0			3	0
		1	0.01			2	0.01
		0	0.01			1	0.01
						0	0.01
		5	0.97			4	0.97
		4	0			3	0
a_2	2	3	0.01	a_6	3	2	0.01
		2	0			1	0.01
		1	0.01			0	0.01
		0	0.01			5	0.96
		4	0.95			4	0
		3	0	a_7	4	3	0.01
a_3	3	2	0.03			2	0.01
		1	0.01			1	0.01
		0	0.01			0	0.01
		5	0.96			6	0.95
		4	0			5	0
a_4	2	3	0.02			4	0.02
		2	0	a_8	4	3	0
		1	0.01			2	0.01
		0	0.01			1	0.01
						0	0.01

vectors is $O(k\delta)$. Removing the duplicates in Step 4 is of order of $O(\delta\log(\delta))$ for all the generated C vectors. The time complexity of Step 5 for checking each vector to be non-minimal is $O(k)$, and as δ is an upper bound for the number of non-duplicate vectors, the time complexity of this step for all the vectors is $O(k\delta)$. Determining the vector X, which corresponds to C, in Step 6 is of order of $O(m)$ for each vector, and accordingly the time complexity of Step 6 for all the vectors is $O(m\delta)$. Similar to Step 4, the time complexity of Step 7 is at most $O(\delta\log(\delta))$ for all the generated (d,τ)-MP^ks. Considering σ as the number of all the obtained (d,τ)-MP^ks, the time complexity of calculating the reliability in Step 7 is $O(m\sigma^2)$ [39]. As a result, the time complexity of Algorithm 3.3 is $O(mk + k\delta + \delta\log(\delta) + m\delta + m\sigma^2)$. As $m > k$, one concludes that the time complexity of the algorithm is $O(m(\delta + \sigma^2))$.

THEOREM 3.8

The time complexity of Algorithm 3.3 is $O(m(\delta + \sigma^2))$.

3.7.4 MORE DISCUSSION

3.7.4.1 Another Algorithm

Algorithm 3.3 applies Theorem 3.6 in Step 5 to remove the non-minimal C vectors and then uses the proposed technique in Ref. [8] in Steps 4 and 7 to remove the duplicate vectors. One can compare all the generated (d,τ)-MP^k candidates to clear the non-minimal and duplicate ones simultaneously. This way, the proposed algorithm in Ref. [18] will be at hand. However, this algorithm is less efficient than Algorithm 3.3 here [10] which shows the reason behind our choice for the algorithm to address the desired problem.

3.7.4.2 Budget Limit

Similar to the previous problems discussed in the preceding sections, one can add some budget limits to this problem. In such a case, one may use Algorithm 3.3 and add a new step to check each obtained (d,τ)-MP^k for budget limit. Alternatively, one may use such a limit to decrease the number of generated *policy vectors* and candidates as it has been done in Ref. [10].

3.8 SOME RELATED OPTIMIZATION PROBLEM

In the last two sections, we discussed the cases with two or more DMPs. It has been supposed that the DMPs are given, and the aim is to assess the network reliability for transmitting a given demand d units of a commodity within a given τ units of time through these paths. Another interesting question raised in these cases is that for a given demand d and time limit τ, which set of disjoint paths is the most reliable one. For the first case, one needs to find a set of two DMPs with the highest reliability index, or in other words, the most reliable set of two DMPs. Similarly, in the other case with more than two DMPs, one needs to determine the most reliable set of k DMPs. As the problems are very similar, we state only an algorithm for the former case.

> **Problem:** What is the most reliable set of two DMPs through which d units of a commodity can be transmitted from the source node to the destination node within τ units of time?
> **Algorithm 3.4** (to address the abovementioned problem)
> **Input**: A multistate flow network with its MPs, time limit τ and demand level d.
> **Output**: The most reliable set of two DMPs.
> **Step 1.** Assuming that P_j, for $j = 1, 2, \ldots, q$ are all the MPs, determine the set of $Q = \{(i,j) \mid P_i$ and P_j are disjoint$\}$. Let $S = \phi$ (set of paths) and $R = 0$ (the current reliability index).
> **Step 2.** For every $(i,j) \in Q$ do
> **2.1.** Evaluate the network reliability, that is, $R_{(d,\tau)}$, by using Algorithm 3.2.
> **2.2.** If $R_{(d,\tau)} > R$, then let $R = R_{(d,\tau)}$ and $S = \{P_i, P_j\}$.

One can easily see that the final set S is the set of most reliable DMPs for which the network reliability is R.

3.9 CONCLUDING REMARKS

Many real-world systems can be modeled as a MFN, and hence reliability evaluation of such networks is of great importance. Several indexes have been introduced in the literature to assess the reliability of a MFN, usually in terms of minimal cuts or paths. They consider some aspects such as budget limit, transmission cost, maintenance cost, transmission time, and time limit. This chapter studied the case with the time limitation and discussed several problems originating from the quickest path problem, including the quickest path reliability problem and its generalization to the cases with two or more DMPs. For each problem, we provided an algorithm, an illustrative example, and the complexity results. Finally, some related optimization problem was introduced and addressed with an algorithm.

ACKNOWLEDGMENT

The author thanks CNPq (grant 306940/2020-5) for supporting this work.

REFERENCES

1. G. Gallo and S. Pallottino. Shortest path algorithms. *Annals of Operations Research*, 13(1):1–79, 1988.
2. Y. L. Chen and Y. H. Chin. The quickest path problem. *Computers & Operations Research*, 17(2):153–161, 1990.
3. Y.-K. Lin. Extend the quickest path problem to the system reliability evaluation for a stochastic-flow network. *Computers & Operations Research*, 30(4):567–575, 2003.
4. C. Yu, J. Lee, and M. J. Munro-Stasiuk. Extensions to least-cost path algorithms for roadway planning. *International Journal of Geographical Information Science*, 17(4):361–376, 2003.

5. M. H. Moore. On the fastest route for convoy-type traffic in flowrate-constrained networks. *Transportation Science*, 10(2):113–124, 1976.

6. G. Ghiani and E. Guerriero. A lower bound for the quickest path problem. *Computers & Operations Research*, 50:154–160, 2014.

7. A. Sedeño-Noda and J. D. González-Barrera. Fast and fine quickest path algorithm. *European Journal of Operational Research*, 238(2):596–606, 2014.

8. M. Forghani-elahabad and N. Mahdavi-Amiri. A new efficient approach to search for all multi-state minimal cuts. *IEEE Transactions on Reliability*, 63(1):154–166, 2014.

9. M. Forghani-elahabad and N. Mahdavi-Amiri. An efficient algorithm for the multi-state two separate minimal paths reliability problem with budget constraint. *Reliability Engineering & System Safety*, 142:472–481, 2015.

10. M. Forghani-elahabad and N. Mahdavi-Amiri. A new algorithm for generating all minimal vectors for the q smps reliability problem with time and budget constraints. *IEEE Transactions on Reliability*, 65(2):828–842, 2015.

11. W.-C. Yeh, W.-W. Chang, and C.-W. Chiu. A simple method for the multi-state quickest path flow network reliability problem. In *2009 8th International Conference on Reliability, Maintainability and Safety*, pages 108–110. IEEE, 2009.

12. W.-C. Yeh and M. J. Zuo. A new subtraction-based algorithm for the d-mps for all d problem. *IEEE Transactions on Reliability*, 68(3):999–1008, 2019.

13. M. Forghani-Elahabad and N. Mahdavi-Amiri. On search for all d-mcs in a network flow. *Iranian Journal of Operations Research*, 4(2):108–126, 2013.

14. M. Forghani-elahabad and N. Mahdavi-Amiri. An improved algorithm for finding all upper boundary points in a stochastic-flow network. *Applied Mathematical Modelling*, 40(4):3221–3229, 2016.

15. Y.-K. Lin and S.-G. Chen. A maximal flow method to search for d-mps in stochastic-flow networks. *Journal of Computational Science*, 22:119–125, 2017.

16. Y.-F. Niu, X.-Z. Xu, C. He, D. Ding, and Z.-Z. Liu. Capacity reliability calculation and sensitivity analysis for a stochastic transport network. *IEEE Access*, 8:133161–133169, 2020.

17. W.-C. Yeh. Fast algorithm for searching d-mps for all possible d. *IEEE Transactions on Reliability*, 67(1):308–315, 2018.

18. Y.-K. Lin. Reliability of k separate minimal paths under both time and budget constraints. *IEEE Transactions on Reliability*, 59(1):183–190, 2010.

19. Y.-K. Lin. Spare routing reliability for a stochastic flow network through two minimal paths under budget constraint. *IEEE Transactions on Reliability*, 59(1):2–10, 2010.

20. M. Forghani-Elahabad, N. Mahdavi-Amiri, and N. Kagan. On multi-state two separate minimal paths reliability problem with time and budget constraints. *International Journal of Operational Research*, 37(4):479–490, 2020.

21. W.-C. Yeh. A fast algorithm for quickest path reliability evaluations in multi-state flow networks. *IEEE Transactions on Reliability*, 64(4):1175–1184, 2015.

22. W.-C. Yeh and M. El Khadiri. A new universal generating function method for solving the single (d, τ)-quick-path problem in multistate flow networks. *IEEE Transactions on Systems, Man, and Cybernetics-Part A: Systems and Humans*, 42(6): 1476–1484, 2012.

23. M. Forghani-elahabad. 1 exact reliability evaluation of multistate flow networks. In *Systems Reliability Engineering*, pages 1–24. De Gruyter, 2021.

24. M. Forghani-elahabad and N. Kagan. An approximate approach for reliability evaluation of a multistate flow network in terms of minimal cuts. *Journal of Computational Science*, 33:61–67, 2019.

25. M. Forghani-elahabad and N. Kagan. Assessing reliability of multistate flow networks under cost constraint in terms of minimal cuts. *International Journal of Reliability, Quality and Safety Engineering*, 26(05):1950025, 2019.

26. S. M. Mansourzadeh, S. H. Nasseri, M. Forghani-elahabad, and A. Ebrahimnejad. A comparative study of different approaches for finding the upper boundary points in stochastic-flow networks. *International Journal of Enterprise Information Systems (IJEIS)*, 10(3):13–23, 2014.

27. Y.-F. Niu, Z.-Y. Gao, and W. H. Lam. Evaluating the reliability of a stochastic distribution network in terms of minimal cuts. *Transportation Research Part E: Logistics and Transportation Review*, 100:75–97, 2017.

28. Y.-F. Niu and X.-Z. Xu. A new solution algorithm for the multistate minimal cut problem. *IEEE Transactions on Reliability*, 69(3):1064–1076, 2019.

29. X. Chen, J. Tao, G. Bai, and Y. Zhang. Search for d-mps without duplications in multistate two-terminal networks. In *2017 Second International Conference on Reliability Systems Engineering (ICRSE)*, pages 1–7. IEEE, 2017.

30. M. Forghani-elahabad and L. H. Bonani. Finding all the lower boundary points in a multistate two-terminal network. *IEEE Transactions on Reliability*, 66(3):677–688, 2017.

31. M. Forghani-elahabad and N. Kagan. Reliability evaluation of a stochastic-flow network in terms of minimal paths with budget constraint. *IISE Transactions*, 51(5): 547–558, 2019.

32. M. Forghani-elahabad and N. Kagan. A simple improved algorithm to find all the lower boundary points in a multiple-node-pair multistate flow network. *Advances in Systems Science and Applications*, 19(1):1–11, 2019.

33. M. Forghani-elahabad, N. Kagan, and N. Mahdavi-Amiri. An mp-based approximation algorithm on reliability evaluation of multistate flow networks. *Reliability Engineering & System Safety*, 191:106566, 2019.

34. Y. Lamalem, K. Housni, and S. Mbarki. An efficient method to find all d-mps in multistate two-terminal networks. *IEEE Access*, 8:205618–205624, 2020.

35. J.-S. Lin, C.-C. Jane, and J. Yuan. On reliability evaluation of a capacitated-flow network in terms of minimal pathsets. *Networks*, 25(3):131–138, 1995.

36. Y.-F. Niu, X.-Y. Wan, X.-Z. Xu, and D. Ding. Finding all multi-state minimal paths of a multi-state flow network via feasible circulations. *Reliability Engineering & System Safety*, 204:107188, 2020.

37. W.-C. Yeh, Z. Hao, M. Forghani-elahabad, G.-G. Wang, and Y.-L. Lin. Novel binary-addition tree algorithm for reliability evaluation of acyclic multistate information networks. *Reliability Engineering & System Safety*, 210:107427, 2021.

38. Z. Zhang and F. Shao. A diameter-constrained approximation algorithm of multistate two-terminal reliability. *IEEE Transactions on Reliability*, 67(3):1249–1260, 2018.

39. A. O. Balan and L. Traldi. Preprocessing minpaths for sum of disjoint products. *IEEE Transactions on Reliability*, 52(3):289–295, 2003.

40. M. Forghani-elahabad and N. Mahdavi-Amiri. An algorithm to search for all minimal cuts in a flow network. *Advances in Systems Science and Applications*, 20(4):1–10, 2020.

4 Optimization of an Innovation Diffusion Model of a High-Valued Product with Delay of Adoption for Low-Income Group

Rakesh Kumar
Department of Applied Sciences, Shaheed Bhagat Singh
State University, Ferozepur, Punjab, India

CONTENTS

4.1 INTRODUCTION

An important aspect in modeling adopter response is to gain information and forecast the impact of marketing strategies. Suppose that marketing manager mails out free samples during the launch of a new product, he wants to know that how many adopters will try the new launch and how this will affect long run market share. But, specifically during a test market or launch of a new product, the dynamical behavior of adopter response is of great important. For example, the managers also wants to know how fast adopters will try the new product and what impact this has on the speed with which the new product achieves its long run share. Moreover, as managers gains more expertise with his market and adapt new strategies as market conditions change. To address these points, an adopter response model should be dynamic and under the controls of marketing strategies and the market environment.

Over the last 60 years, a variety of marketing systems have been developed in which the prices of the products and/or the advertisements are control variables whose optimum values must be gained over a given planning period of length T. Such type of models are more relevant during the launching phase of a new innovation (product) where the prices as well as advertisements changes can affect more or less strongly the "new adopters" and "imitators" in the social system of potential buyers. The most important and dramatic aspects of the "learning curve" depicting production costs during this period, which should also be taken into consideration when optimum price and advertisements strategies are to be determined. The important papers which involve such important systems are: Bass [1], Bass and Bultez [2], Doland and Jeuland [3], Dorfman and Steiner [4], Erickson [5], Gensch and Whelam [6, 7], Horsky and Simon [8], Kalish [9], Kalish and Lilien [10], Kotowitz and Mathews on [11], Robinson and Lakhani [12], Sethi [13], and Teng and Thompson [14, 15]. A lengthy survey of these kinds of models was given by Sethi [16]. Each of these papers contains a theoretical treatment of its particular model, and, as might be expected, there is considerable duplication and overlap in these treatments. A more exhaustive survey article would be required to go into comprehensive comparisons of these models. However, the author will be focused to discuss the role of delay parameter in the adoption of the new product. It is worth mentioning the work of many mathematicians who worked in the area of innovation diffusion modeling by considering time delay parameter in their systems, and justified the explanations and role of delay in the field of innovation diffusion modeling: Kumar and Sharma [17, 18].

The main aim of the chapter is to gain a general framework for the theoretical treatment of price-advertising models which contain as special cases all of the models mentioned in the previous paragraph as well as others yet to be devised.

This chapter is organized as follows: In Section 4.1, introduction of the problem is given, Section 4.2, a nonlinear innovation diffusion model incorporating time delay as control strategy, external as well as internal influences are presented. In Section 4.3, the positivity of the solutions and boundedness of the system is checked. In Section 4.4, the dynamical behavior of the time delayed model is studied and obtained the conditions for local asymptotic stability in case of adopter-free equilibrium and interior equilibrium. The system has also been examined for Hopf bifurcation and obtained the existence conditions for Hopf bifurcation by using bifurcation theory in Section 4.5. In Section 4.6, some numerical simulations have been presented to support the analytical findings. Finally, a precise conclusion has been produced in the concluding section.

4.2 THE MODEL FORMULATION

In this section, the author focused on to formulate a general cost-internal influences model and explain suppositions which are appropriate for such kind of model. Thereafter, the author characterized the optimal internal influences (word-of-mouths) strategies for the model system.

Advertisements are the most commonly used and important strategic element in finding out the market share achieved by a firm in different marketing situations. The cost is another important element in a marketing strategy which affects both the demand for and the profit share obtained from a product. Also, the demand of a product is determined by the cumulative density of adopters. Another important aspect is the word-of-mouth influences. Previous adopters transmit knowledge about the product, thus remove the uncertainties linked with it, and increase the demand for the product. That is why author will consider the growth of the demand for a product as a dynamic function of internal factors (word-of-mouth interactions).

The innovation diffusion model system is formulated by taking into consideration the cost of a product. The products that are high end (costlier) are not a cup of tea for middle-income group population in the region instantaneously. The population has been categories as the non-adopters population of high-income group with population density $N_1(t)$, the non-adopters population of middle class group with density $N_2(t)$, and the adopters with population density $A(t)$, where t denotes the time. To make it more effective and realistic, the various demographic processes of the population have been taken into consideration for the formulation of the model. The basic assumptions are as below:

1. Suppose that r_1 and r_2 are the new recruitment rates of non-adopter populations $N_1(t)$ and $N_2(t)$ respectively.
2. The financial status of the people may be varied with the passage of time. It means that the non-adopter population of middle-income group $N_1(t)$ can be the population of high-income group $N_2(t)$ in future times and vice-versa. It means that their populations can migrate to the groups of each other according to their current financial positions. Assume that ξ_1 is the rate at which the non-adopter population of high-income group $N_1(t)$ join the population of middle class income group $N_2(t)$ and ξ_2 is the rate at which the non-adopter population of middle class income group $N_2(t)$ join the population of high-income group $N_1(t)$.
3. As we are taking the high-valued innovations, so it is supposed that the non-adopters population of high-income group $N_1(t)$ may join the adopter's class instantaneously to adopt the product.
4. Suppose that the variable internal influences (word of mouth) directly affect the decision about the adoption. Therefore, the rate of change of the adopters depends on internal influences.
5. Let α be the rate of mutual interactions (word of mouth) occurred between the non-adopter population $N_1(t)$ and the adopter population $A(t)$. So $\alpha N_1(t)A(t)$ is the non-adopters directly transferring to adopter population class $A(t)$ who are having information about the product at time t by virtue of internal influences.
6. Also, assume that the middle class population takes some average time for evaluating the product, and then to make final decision about the adoption of the product, i.e., the shifting of non-adopters $N_2(t)$ to the adopter class $A(t)$ is not instantaneous, but takes small time defined by the average evaluation period τ. Moreover, the individuals who have knowledge of the product at time $t-\tau$ may leave the evaluation stage in the interval $[t-\tau,t]$. Let β be the rate at which cumulative density of mutual interactions (variable internal

influences) occurred in the interval $[t-\tau,t]$, to affect the decisions of the non-adopters population $N_2(t)$. So $\beta N_2(t-\tau)A(t)$ is the non-adopters directly transferring to adopter population having information about the product at time $t-\tau$ by internal influences, who join the adopter's class $A(t)$ at time t.

7. Let v_1 and v_2 be the rates at which the adopters joining back to the non-adopter population class $N_1(t)$ and $N_2(t)$ respectively due to dissatisfaction by the performance of the product. Suppose that μ is the parameter used for the death and emigration rates of the various population classes $N_1(t)$, $N_2(t)$ and $A(t)$.

Thus, the mathematical form of the model is given as below:

$$\frac{dN_1}{dt} = r_1 - \alpha N_1(t)A(t) - \xi_1 N_1(t) + \xi_2 N_2(t) + v_1 A(t) - \mu N_1(t),$$

$$\frac{dN_2}{dt} = r_2 + \xi_1 N_1(t) - \xi_2 N_2(t) - \beta N_2(t-\tau)A(t) + v_2 A(t) - \mu N_2(t), \qquad (4.1)$$

$$\frac{dA}{dt} = \alpha N_1(t)A(t) + \beta N_2(t-\tau)A(t) - (\mu + v_1 + v_2)A(t).$$

These equations are to be solved subject to the initial conditions

$$N_1(\theta) = \phi_1(\theta), N_2(\theta) = \phi_2(\theta), A(\theta) = \phi_3(\theta); \quad \phi_i(0) > 0, \forall\, i = 1,2,3 \qquad (4.2)$$

The functions $\phi_1(\theta), \phi_2(\theta), \phi_3(\theta)$ are continuous functions which are bounded in the interval $[-\tau,0]$. Specifically, $\phi_1(\theta), \phi_2(\theta), \phi_3(\theta) \in C([-\tau,0], \mathfrak{R}_+^3)$, the Banach space of functions mapping the interval $[-\tau,0]$ into $\mathfrak{R}_+^3 = \{(y_1, y_2, y_3): y_i > 0, \quad i = 1,2,3\}$, which are continuous. Applying the basic results of FDE [19], make a note that all the solutions $N_1(\theta), N_2(\theta), A(\theta)$ of the IVP are always unique and nonnegative on $[0,+\infty)$.

The descriptions of the various notations and symbols used in the chapter is given in following Table 4.1

TABLE 4.1
Parameters Used with Their Meanings

Symbol	Usual Meaning
$N_1(t)$	Non-adopters population density of high-income group
$N_2(t)$	Non-adopters population density of middle-income group
$A(t)$	Adopters population density
r_1	Recruitment rates of non-adopter populations $N_1(t)$
r_2	Recruitment rates of non-adopter populations $N_2(t)$
ξ_1	Transfer rate at which $N_1(t)$ join $N_2(t)$
ξ_2	Transfer rate at which $N_2(t)$ join $N_1(t)$
α	Rate of mutual interactions between the non-adopter population $N_1(t)$ and the adopter population $A(t)$
τ	Average evaluation time period
β	Rate of mutual interactions occurred between $N_2(t)$ and $A(t)$ in $[t-\tau,t]$
v_1	Rate at which the adopters $A(t)$ joining back the non-adopter population class $N_1(t)$
v_2	Rate at which the adopters $A(t)$ joining back the non-adopter population class $N_2(t)$
μ	Natural death and emigration rates of all the population classes

4.3 BASIC RESULTS

For the well-posed of the system, there should exist positive solutions of model (4.1) lying in the boundary of a region. So, the author presents the following theorems for proving the boundedness and positivity of the model system (4.1).

THEOREM 4.1 [18]

All solutions of model (4.1) subject to the initial values (4.2) are non-negative, $\forall\, t \geq 0$.

Proof. For $t \in [0, \tau]$, the first equation of model (4.1) can be rewritten as

$$\frac{dN_1}{dt} \geq -(\alpha A(t) + \xi_1 + \mu)N_1(t),$$

This gives $N_1(t) \geq N_1(0)\exp\left[-\int_0^t (\alpha A(v) + \xi_1 + \mu)dv\right] > 0.$

The second equation of the system can also be written as

$$\frac{dN_2}{dt} \geq -(\xi_2 + \mu)N_2(t) - \beta N_2(t - \tau)A(t)$$

$$\frac{dN_2}{dt} + (\xi_2 + \mu)N_2(t) \geq -\beta N_2(t - \tau)A(t)$$

On solving, one can get

$$N_2(t) \geq e^{-\int_0^t (\xi_2 + \mu)ds}\left[N_2(0) + \int_0^t \left\{\beta N_2(s - \tau)A(s)\right\}e^{\int_0^s (\xi_2 + \mu)d\theta}\, ds\right] > 0, \quad t \in [0, \tau].$$

The third equation of the system (4.1) gives

$$\frac{dA}{dt} + (\mu + v_1 + v_2 - \alpha N_1)A(t) \geq \beta N_2(t - \tau)A(t),$$

Solving the equation, we shall have

$$A(t) \geq e^{-\int_0^t (\mu + v_1 + v_2 - \alpha N_1(s))ds}\left[A(0) + \int_0^t \left\{\beta N_2(s - \tau)A(s)\right\}e^{\int_0^s (\mu + v_1 + v_2 - \alpha N_1(\theta))d\theta}\, ds\right] > 0,$$

for $t \in [0, \tau]$.

Similarly, one can take the successive intervals $[\tau, 2\tau], [2\tau, 3\tau], \ldots$ $[n\tau, (n+1)\tau]; n \in \mathbb{N}$ [20] and make a result that all the solutions of model (4.1) always positive $\forall\, t \geq 0$.

LEMMA 4.1 [21]

The positive solutions of the system (4.1) with initial values (4.2) in $\mathfrak{R}_+^3$ are uniformly bounded.

Proof. Suppose that a function $\Sigma(t) = N_1(t) + N_2(t) + A(t), \forall t \geq 0$. Differentiating w.r.t. t, and obtain

$$\frac{d\Sigma}{dt} = r_1 + r_2 - \mu(N_1(t) + N_2(t) + A(t)),$$

$$\frac{d\Sigma}{dt} = r_1 + r_2 - \mu\Sigma,$$

or $\dfrac{d\Sigma}{dt} + \mu\Sigma = r_1 + r_2$, for any $\mu > 0$.

Using an important result on differential inequalities, one can obtain

$$0 \leq \Sigma(t) \leq \frac{r_1 + r_2}{\mu}(1 - e^{-\mu t}) + \Sigma(0)e^{-\mu t}, \text{ for any } \mu > 0, \qquad (4.3)$$

As a limiting form $t \to \infty$, equation (4.3) gives $0 \leq \Sigma(N_1, N_2, A) \leq \dfrac{r_1 + r_2}{\mu}$.

Thus, all solutions of model (4.1) which start in $\mathfrak{R}_+^3$ will always be in the region $\Psi = \{(N_1(t), N_2(t), A(t)) : 0 \leq \Sigma(t) \leq \dfrac{r_1 + r_2}{\mu} + \varepsilon, \text{for any } \varepsilon > 0\}$. Thus $N_1(t), N_2(t), A(t)$ will always have upper bound in the region $\Psi, \forall t \geq 0$.

4.4 EQUILIBRIUM POINTS

The system (4.1) will have two equilibrium points:

i. $U^0 = (N_1(t), N_2(t), 0)$, which is an adopter free equilibrium point, which is given by

$$N_1(t) = \frac{(\xi_1 + r_2)\xi_2 + (\xi_2 + \mu)r_1}{(\xi_1 + \mu)(\xi_2 + \mu)}, N_2(t) = \frac{\xi_1 + r_2}{\xi_2 + \mu}.$$

ii. $U^* = (N_1^*(t), N_2^*(t), A^*(t))$ is the interior equilibrium point.

Furthermore, the solutions of the subsequent set of mathematical equations provide the interior equilibrium point $U^* = (N_1^*(t), N_2^*(t), A^*(t))$ of the proposed model system (4.1):

$$r_1 - \alpha N_1(t)A(t) - \xi_1 N_1(t) + \xi_2 N_2(t) + v_2 A(t) - \mu N_1(t) = 0,$$
$$r_2 + \xi_1 N_1(t) - \xi_2 N_2(t) - \beta N_2(t)A(t) + v_2 A(t) - \mu N_2(t) = 0, \qquad (4.4)$$
$$\alpha N_1(t)A(t) + \beta N_2(t)A(t) - (\mu + v_1 + v_2)A(t) = 0,$$

From first equation of system Equation (4.4), one can obtain

$$N_1^* = \frac{\mu + v_1 + v_2 - \beta N_2^*}{\alpha}.$$

Using the value of N_1^* in second equation of (4.4), and for the positive existence of N_2^*, one should have

$$\alpha > \beta \text{ and } (r_1 + r_2)v_2 + \mu r_2 + \frac{\mu \xi_1(\mu + r_1 + r_2)}{\alpha} < \frac{\mu v_2(\mu + r_1 + r_2)}{\alpha}.$$

Also, for the positivity of A^*,

$$\frac{v_2}{\beta} < N_2^* < \text{Min}\left\{ \frac{\mu + v_1 + v_2}{\beta}, \frac{r_2 + \frac{\xi_1(\mu + v_1 + v_2)}{\alpha}}{\frac{\xi_1 \beta}{\alpha} + \xi_2 + \mu} \right\}.$$

So, for the positive existence of interior equilibrium point $U^* = (N_1^*(t), N_2^*(t), A^*(t))$, the following theorem can be stated:

THEOREM 4.2

The model system (4.1) has an interior equilibrium point if

1. $\mu + v_1 + v_2 > \dfrac{\alpha}{\beta N_2^*}$,

2. $2\alpha > \beta$ and $(r_1 + r_2)v_2 + \mu r_2 + \dfrac{\mu \xi_1(\mu + r_1 + r_2)}{\alpha} < \dfrac{\mu v_2(\mu + r_1 + r_2)}{\alpha}$,

3. $\dfrac{v_2}{\beta} < N_2^* < \text{Min}\left\{ \dfrac{\mu + v_1 + v_2}{\beta}, \dfrac{r_2 + \frac{\xi_1(\mu + v_1 + v_2)}{\alpha}}{\frac{\xi_1 \beta}{\alpha} + \xi_2 + \mu} \right\}., \forall\, t \geq 0.$

4.5 DYNAMICAL ANALYSIS OF THE SYSTEM

In the current section, the author proceeds to examine the asymptotic stability of model (4.1) for $\tau > 0$ and the Hopf-bifurcation analysis of the system about $U^* = (N_1^*(t), N_2^*(t), A^*(t))$ and, hence, will be validated by taking a suitable example in the numerical simulations section.

4.5.1 STABILITY OF THE EQUILIBRIUM POINT U^0

The variational matrix of the system (4.1) at $U^0 = (N_1(t), N_2(t), 0)$ is given by

$$J^0 = \begin{bmatrix} -\xi_1 - \mu - \lambda & \xi_2 & v_1 \\ \xi_1 & -\xi_2 - \mu - \lambda & -\beta N_2^* + v_2 \\ 0 & 0 & \alpha N_1^* + \beta N_2^* - (\mu + v_1 + v_2) - \lambda \end{bmatrix}$$

The characteristic equation for the equilibrium point $U^0 = (N_1(t), N_2(t), 0)$ can be obtained by solving the determinant $|J^0 - \lambda I| = 0$ and can be detailed as below:

$$-(\xi_1 + \mu + \lambda)(\xi_2 + \mu + \lambda)(\lambda - \alpha N_1^* - \beta N_2^* + (\mu + v_1 + v_2))$$
$$-\xi_1 \xi_2 (\alpha N_1^* + \beta N_2^* - (\mu + v_1 + v_2) - \lambda) = 0. \tag{4.5}$$

On solving (4.5), one can obtain

$$(\lambda + (\mu + v_1 + v_2) - \alpha N_1^* - \beta N_2^*)((\xi_1 + \mu + \lambda)(\xi_2 + \mu + \lambda) - \xi_1 \xi_2) = 0,$$

$$(\lambda + (\mu + v_1 + v_2) - \alpha N_1^* - \beta N_2^*)((\lambda + \mu)^2 + (\lambda + \mu)\xi_1 + (\lambda + \mu)\xi_2) = 0,$$

$$(\lambda + \mu)^2 + (\lambda + \mu + \xi_1 + \xi_2)(\lambda + (\mu + v_1 + v_2) - \alpha N_1^* - \beta N_2^*) = 0. \tag{4.6}$$

The eigenvalues are given by Equation (4.6)

$$\lambda = -\mu, -(\mu + \xi_1 + \xi_2), \alpha N_1^* + \beta N_2^* - (\mu + v_1 + v_2).$$

After observing all the eigenvalues, it has been noticed that the first two eigenvalues are purely negative, and the third value is conditionally negative. So, the system will locally stable at $U^0 = (N_1(t), N_2(t), 0)$ if

$$\alpha N_1^* + \beta N_2^* - (\mu + v_1 + v_2) < 0$$

$$\text{or if } \alpha \left(\frac{(\xi_1 + r_2)\xi_2 + (\xi_2 + \mu)r_1}{(\xi_1 + \mu)(\xi_2 + \mu)} \right) + \beta \left(\frac{\xi_1 + r_2}{\xi_2 + \mu} \right) < (\mu + v_1 + v_2) \tag{4.7}$$

$$\text{or if } \frac{\alpha \xi_2 (\xi_1 + r_2) + \alpha r_1 (\xi_2 + \mu) + \beta(\xi_1 + r_2)(\xi_2 + \mu)}{(\xi_1 + \mu)(\xi_2 + \mu)} < (\mu + v_1 + v_2).$$

4.5.2 STABILITY OF THE EQUILIBRIUM POINT U^*

For the delayed innovation diffusion model (4.1), the Jacobian matrix obtained about the interior point U^* is given as:

$$J^* = \begin{bmatrix} -\alpha A^* - \xi_1 - \mu - \lambda & \xi_2 & v_1 \\ \xi_1 & -\xi_2 - \beta A^* e^{-\lambda \tau} - \mu - \lambda & -\beta N_2^* + v_2 \\ \alpha A^* & \beta A^* e^{-\lambda \tau} & \alpha N_1^* + \beta N_2^* - (\mu + v_1 + v_2) - \lambda \end{bmatrix}$$

The characteristic equation for J^* at U^* is obtained by taking the determinant $|J^* - \lambda I| = 0$, as

$$(\lambda^3 + A_1 \lambda^2 + A_2 \lambda + A_3) + (B_1 \lambda^2 + B_2 \lambda + B_3) e^{-\lambda \tau} = 0, \tag{4.8}$$

where

$A_1 = \alpha A^* + \xi_1 + \xi_2 + 2\mu - \alpha N_1^* + \beta N_2^* + (\mu + v_1 + v_2);$

$A_2 = (\xi_2 + \mu)\big((\mu + v_1 + v_2) - \alpha N_1^* - \beta N_2^*\big) + (\alpha A^* + \xi_1 + \mu)\big(\xi_2 + \mu + (\mu + v_1 + v_2) - \alpha N_1^* - \beta N_2^*\big)$

$\quad - \xi_1 \xi_2 - \alpha v_1 A^*;$

$A_3 = (\alpha A^* + \xi_1 + \mu)(\xi_2 + \mu)\big((\mu + v_1 + v_2) - \alpha N_1^* - \beta N_2^*\big) - \xi_1 \xi_2 \big((\mu + v_1 + v_2) - \alpha N_1^* - \beta N_2^*\big)$

$\quad - \alpha A^* \left(v_2 - \beta N_2^*\right);$

$B_1 = \beta A^*;$

$B_2 = \beta A^* \left(\alpha A^* + \xi_1 + \mu + (\mu + v_1 + v_2) - \alpha N_1^* - \beta N_2^*\right);$

$B_3 = \beta A^* \left(\alpha A^* + \xi_1 + \mu\right)\big((\mu + v_1 + v_2) - \alpha N_1^* - \beta N_2^*\big) - \xi_1 v_1 \beta A^* - -\xi_2 \alpha \beta v_1 A^{*2};$

Let us check for the stability analysis of steady state U^* of model (4.1). Firstly, let us analyze the model in the absence of τ. Without parameter τ, Equation (4.4) becomes

$$\lambda^3 + (A_1 + B_1)\lambda^2 + (A_2 + B_2)\lambda + (A_3 + B_3) = 0, \qquad (4.9)$$

With the help of Routh-Hurwitz theorem, all values of Equation (4.9) will be negative, i.e., the interior steady state U^* is LAS (locally asymptotically stable) provided (S_1) holds good, where

$$S_1 : (i)\ A_1 + B_1 > 0\ (ii)\ A_2 + B_2 > 0, (iii)\ A_3 + B_3 > 0,$$

$$\text{and } (iv)(A_1 + B_1)(A_2 + B_2) - (A_3 + B_3) > 0. \qquad (4.10)$$

4.5.3 HOPF BIFURCATION AT EQUILIBRIUM POINT U^*

For the Hopf-bifurcation analysis, the following lemma should be in our state of mind:

LEMMA 4.2 [22]

i. The point U^* of (4.1) is absolutely stable if the point U^* of the corresponding ODE model is locally asymptotically stable and the transcendental characteristic Equation (4.4) will have real roots for $\tau > 0$.

ii. The equilibrium U^* of (4.1) is conditionally stable if all roots of (4.4) have negative real parts for zero τ, and $\exists\ \tau > 0$ so that Equation (4.4) has an imaginary eigenvalues of the form $\pm i\omega$.

The following theorem may also be stated:

THEOREM 4.3 [23]

The interior point U^* is conditionally stable if (S_1) hold good for model (4.1).

To check how the parameter τ (time delay) affects the conditions of the stability of positive steady state U^*, let τ be the bifurcation parameter. To find the instability incurred in the system (4.1) by a delay parameter τ, let $\lambda = i\omega$, for any $\omega > 0$, be a root of (4.4) ($\tau > 0$). Put $\lambda = i\omega$ into Equation (4.4), and find out real part and imaginary part of the characteristic Equation (4.4),

$$B_2\omega \sin(\omega\tau) + (B_3 - B_1\omega^2)\cos(\omega\tau) = A_1\omega^2 - A_3, \qquad (4.11)$$

$$B_2\omega \cos(\omega\tau) + (B_3 - B_1\omega^2)\sin(\omega\tau) = \omega^3 - A_2\omega. \qquad (4.12)$$

Solving Equations (4.11) and (4.12) for sine and cosine terms, and using the formula of trigonometry $\sin^2 \psi + \cos^2 \psi = 1$, one can easily get after simple computations

$$\Theta(s) = s^3 + \Phi_1 s^2 + \Phi_2 s + \Phi_3 = 0, \qquad (4.13)$$

$$\text{where} \begin{cases} s = \omega^2, \text{ and } \Phi_1 = P_1^2 - 2P_2 - Q_1^2, \\ \Phi_2 = P_2^2 - 2P_1P_3 + 2Q_1Q_3 - Q_2^2, \quad \Phi_3 = P_3^2 - Q_3^2 \end{cases} \qquad (4.14)$$

If there exists a minimum of one nonnegative root of equation, the characteristic Equation (4.8) will have purely imaginary roots. So, $\omega = \pm\sqrt{s_1}$ will be the pair of eigenvalues, which are complex conjugate, and this is possible if s_1 (say) is a positive real root of (4.13). Thus, the Hopf bifurcation might exist in the system (4.1).

The subsequent theorem explains the possibility of the existence of a positive real root of (4.13):

THEOREM 4.4 [22]

Equation (4.13) $\Theta(s) = 0$ has a minimum one positive real root if one of the given conditions are justified:

(i) $\Phi_3 < 0$, (ii) $\Phi_3 \geq 0$, $\Phi_1^2 - 3\Phi_2 > 0$, and $s_c > 0$ of $\Theta(s)$ exists with $\Theta(s_c) \leq 0$.

Thus, the model system (4.1) has purely imaginary eigenvalues if and only if conditions (i) or (ii) in Lemma 4.2 are verified. Also Equation (4.13) can have at most three positive real roots, $s_i > 0$; $i = 1, 2, 3$, it is also possible that there may exist three purely imaginary pairs of eigenvalues, $\lambda_i = i\omega_i = \pm i\sqrt{s_i}$, $i = 1, 2, 3$. Let us now find the values of the time delay parameter τ_i corresponding to values of ω_i, and by applying ω_i into Equations (4.11) and (4.12) for the real and imaginary parts of the characteristic Equation (4.8). Solving for $\sin(\omega_i\tau)$ and $\cos(\omega_i\tau)$, one can easily obtain the following:

$$\cos(\omega_i\tau) = \frac{(\omega_i^3 - A_2\omega_i)B_2\omega_i - (A_1\omega_i^2 - A_3)(B_3 - B_1\omega_i^2)}{B_2^2\omega_i^2 - (B_3 - B_1\omega_i^2)^2}, \quad (4.15)$$

$$\sin(\omega_i\tau) = \frac{(A_1\omega_i^2 - A_3)B_2\omega_i - (\omega_i^3 - A_2\omega_i)(B_3 - B_1\omega_i^2)}{B_2^2\omega_i^2 - (B_3 - B_1\omega_i^2)^2}, \quad (4.16)$$

The thresh value of τ at which the system exhibits stability switch, i.e., values of (4.8) at which complex roots occur, can be obtained by using Equations (4.15) and get that

$$\tau_i^{(j)} = \frac{1}{\omega_i}\arccos\left(\frac{(\omega_i^3 - A_2\omega_i)B_2\omega_i - (A_1\omega_i^2 - A_3)(B_3 - B_1\omega_i^2)}{B_2^2\omega_i^2 - (B_3 - B_1\omega_i^2)^2}\right) + \frac{2j\pi}{\omega_i}, \quad (4.17)$$

where $i = 0, 1, 2$, and $j = 0, 1, 2, 3...$ The least value of τ_0 at which the purely imaginary eigenvalues of the form $\lambda_0 = \pm i\omega_0$ occur is therefore given as below:

$$\tau_0 = \min_{0 \le i \le 2, j \ge 0} \tau_i^{(j)}, \quad \tau_i^{(j)} > 0.$$

Now, let us find the condition of Hopf bifurcation for the system (4.1) around the interior point U^*. Taking τ as a parameter of bifurcation and suppose that $\lambda = \zeta + i\omega$ ($\tau > 0$) is a root of Equation (4.8), where $\omega > 0$ is a real. Putting $\lambda = \zeta + i\omega$ into (4.8) and obtained as follows:

$$\zeta^3 - 3\mu\omega^2 + A_1(\zeta^2 - \omega^2) + A_2\zeta + A_3$$
$$+ \left[\left\{B_1(\zeta^2 - \omega^2) + B_2\zeta + B_3\right\}\cos(\omega\tau) + (2B_1\zeta\omega + B_2\omega)\sin(\omega\tau)\right]e^{-\zeta\tau} = 0, \quad (4.18)$$

and $(-\omega^3 + 3\zeta^2\omega + 2A_1\zeta\omega + A_2\omega)$
$$+ \left[-\left\{B_1(\zeta^2 - \omega^2) + B_2\zeta + B_3\right\}\sin(\omega\tau) + (2B_1\zeta\omega + B_2\omega)\cos(\omega\tau)\right]e^{-\zeta\tau} = 0, \quad (4.19)$$

Differentiating Equations (4.18) and (4.19) w.r.t. τ and letting $\tau = \hat{\tau}, \omega = \hat{\omega}$, and $\zeta = 0$, the obtained expressions are as below:

$$G_1\left[\frac{d\zeta}{d\tau}\right]_{\tau=\hat{\tau}} - G_2\left[\frac{d\omega}{d\tau}\right]_{\tau=\hat{\tau}} = H_1, \quad (4.20)$$

$$G_2\left[\frac{d\zeta}{d\tau}\right]_{\tau=\hat{\tau}} + G_1\left[\frac{d\omega}{d\tau}\right]_{\tau=\hat{\tau}} = H_2, \tag{4.21}$$

where

$$\begin{cases} G_1 = -3\hat{\omega}^2 + A_2 + \left\{B_2 + \tau(B_1\omega^2 - B_3)\right\}\cos(\omega\tau) + \omega(2B_1 - B_2\tau)\sin(\omega\tau), \\ G_2 = 2A_1\omega - \left\{B_2 + \tau(B_1\omega^2 - B_3)\right\}\sin(\omega\tau) + +\omega(2B_1 - B_2\tau)\cos(\omega\tau), \\ H_1 = \omega(B_3 - B_1\omega^2)\sin(\omega\tau) - B_2\omega^2\cos(\omega\tau), \\ H_2 = \omega(B_3 - B_1\omega^2)\cos(\omega\tau) + B_2\omega^2\sin(\omega\tau). \end{cases} \tag{4.22}$$

Simplify Equations (4.20) and (4.21), and at $\tau = \hat{\tau} = \tau_0, \omega = \hat{\omega}^2 = \omega_0^2$,

$$\left[\frac{d\zeta}{d\tau}\right]_{\tau=\tau_0} = \frac{\hat{\omega}^2}{G_1^2 + G_2^2}\left[\frac{d\Theta}{ds}\right]_{\omega=\omega_0^2} \neq 0. \tag{4.23}$$

Hence, the values of the transcendental Equation (4.8) cross the vertical axis as the bifurcation parameter τ crosses over the threshold value. Hence, at $\tau = \tau_0$, which is the smallest positive value of τ given by Equation (4.17), the conditions for Hopf bifurcation are justified.

THEOREM 4.5 [24]

Suppose that $U^* = (N_1^*(t), N_2^*(t), A^*(t))$ exist and the condition in (S_1) are satisfied for the innovation diffusion model (4.1), then the conditions for $U^* = (N_1^*(t), N_2^*(t), A^*(t))$ to be LAS with τ are

  i. if $\tau \in [0, \tau_0)$, U^* is LAS;
  ii. if $\tau \geq \tau_0$, the point U^* bifurcates into periodic orbits, i.e., it becomes unstable;
  iii. system (4.1) undergoes Hopf bifurcation at threshold value τ_0 around U^* where

$$\tau_0 = \frac{1}{\omega_0}\arccos\left(\frac{(\omega_0^3 - A_2\omega_0)B_2\omega_0 - (A_1\omega_0^2 - A_3)(B_3 - B_1\omega_0^2)}{B_2^2\omega_0^2 - (B_3 - B_1\omega_0^2)^2}\right). \tag{4.24}$$

4.6 NUMERICAL SIMULATIONS

Here, the results that are obtained analytically in the previous sections will be validated by the help of a numerical example and satisfy the outcomes of the given model system (4.1). So, let us suppose a hypothetical set of parametric values which are exhibited in the following example as

$$\frac{dN_1}{dt} = 0.25 - 0.2N_1(t)A(t) - 0.3N_1(t) + 0.31N_2(t) + 0.08A(t) - 0.2N_1(t),$$

$$\frac{dN_2}{dt} = 0.24 + 0.3N_1(t) - 0.31N_2(t) - 0.4N_2(t-\tau)A(t) + 0.01A(t) - 0.2N_2(t), \quad (4.25)$$

$$\frac{dA}{dt} = 0.2N_1(t)A(t) + 0.4N_2(t-\tau)A(t) - (0.2 + 0.08 + 0.01)A(t).$$

The exhaustive explanations of the numerical example can be represented by considering subsequent cases:

Case I: $\tau = 0$
The Matlab software has been used to integrate model (4.25). By taking different sets of initial values, the system (4.25) observed locally asymptotically stable behavior in the absence of time delay parameter τ, and the system (4.25) converges to a stable steady state $U^0(1.082, 0.3708, 1.799)$. It has been depicted by the help of Figure 4.1

Furthermore, the conditions (4.10) for local asymptotic stability of system (4.1)

$$S_1 : A_1 + B_1 = 1.0152 > 0, \; A_2 + B_2 = 0.618 > 0, \; A_3 + B_3 = 0.216 > 0,$$
$$(A_1 + B_1)(A_2 + B_2) - (A_3 + B_3) = 0.4336 > 0$$

have also been satisfied numerically.

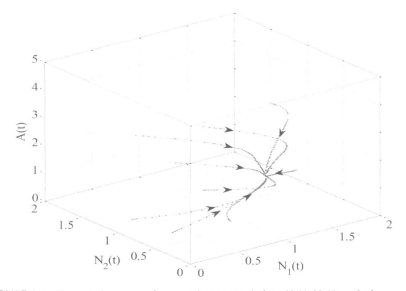

FIGURE 4.1   The solution curves for non-adopter populations $N_1(t), N_2(t)$, and adopter population $A(t)$ are converging to stable equilibrium $U^0(1.082, 0.3708, 1.799)$. It means that the populations are converging to stable equilibrium position without any delay.

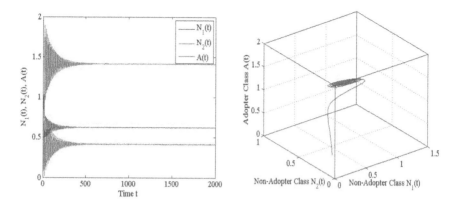

FIGURE 4.2 The solution curves for Non-adopter and Adopter classes of the system (4.18) are converging to $U^*(0.6238, 0.4132, 1.402)$ at $\tau = 3.4 < 3.5 = \tau_0$. The populations are producing stable dynamics for $\tau = 3.4$.

Case II: $\tau > 0$,

The system (4.25) has again been integrated with initial values (0.3, 0.3, 0.1), and for $\tau > 0$ and observed that there exists a purely imaginary root $i\omega$, where $\omega = 3.816$ has been calculated from (4.14), and corresponding to this value, the critical value of time delay parameter $\tau_0 = 3.5$ is calculated from Equation (4.17), so that, the positive steady state U^* remains stable for $0 \le \tau < 3.5$ and remains unstable for $\tau \ge 3.5$ (Figures 4.2 and 4.3). Moreover, the transversality condition from Equation (4.23), i.e., $\left[\dfrac{d\zeta}{d\tau} \right]_{\tau=\tau_0, \omega=\omega_0} = 10.2816 \ne 0$ is also verified, which is

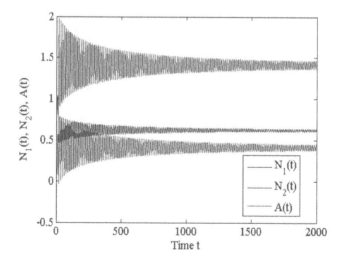

FIGURE 4.3 Time series produces periodic solution trajectories appears around the positive equilibrium $U^*(0.6197, 0.4031, 1.421)$ at $\tau = 3.5$. The population curves are bifurcating into small periodic orbits.

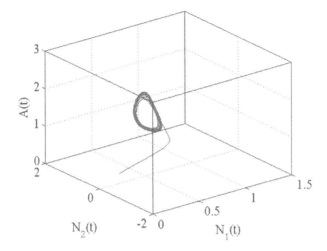

FIGURE 4.4 Phase plane depicts more stable periodic solutions around the positive equilibrium $U^*(0.5739, 0.2345, 1.638)$ at $\tau = 4.3$. All the population curves are justifying Hopf bifurcation around the positive equilibrium U^*.

the prior condition for the existence of Hopf bifurcation. Thus, the system (4.1) produces Hopf bifurcation via a limit cycle and has been presented in the form of a more stable limit cycle in Figure 4.4 for $\tau = 4.3$. Thus, it proves that there exists a threshold limit of delay parameter τ beyond which, the system shows excitability in the form of limit cycle. The dynamical behavior of model (4.1) with the occurrence of a Hopf bifurcation has also been presented in Figure 4.5, where the bifurcation diagram of model (4.1) with respect to the bifurcating parameter τ has been shown. The analytical and numerical simulations demonstrated that the parameter τ in the innovation diffusion process of (4.1) has proved the bifurcating periodic orbits in the dynamics of the system. The outcomes proved that the time delay parameter is must for the innovation to take off in various markets, and it has destabilizing effects on the innovation diffusion model (4.1). It has been transforming the system (4.1) from a locally stable behavior to a limit cycle via Hopf bifurcation.

4.6.1 Role of Internal Factors in the Optimization Process

Here, the role of internal factors (word of mouth interactions) has been observed in optimizing the innovation diffusion process of an innovation (high-valued product). It has been investigated that when internal factors in the innovation diffusion have been increased from $\beta = 0.4$ to 0.7, the system (4.1) depicted irregularities (2-period, 3-period periodic orbits) in the system, i.e., the unpredictable situations have been arisen, and it has been shown in Figure 4.6.

It means that for the optimization of innovation diffusion process of a high-valued product, the optimality of word-of-mouth is quite needed.

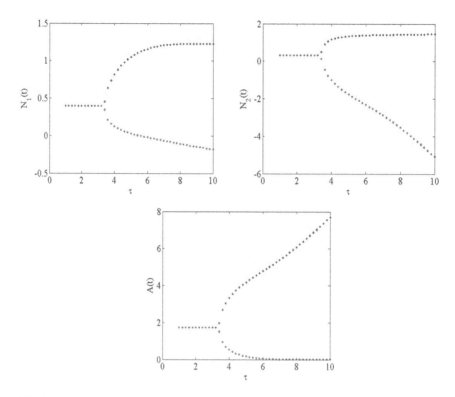

FIGURE 4.5 The bifurcation plot of model (4.1) w.r.t. the bifurcating parameter τ. The diagram is clearly depicting the existence of a stable limit cycle around the interior equilibrium point.

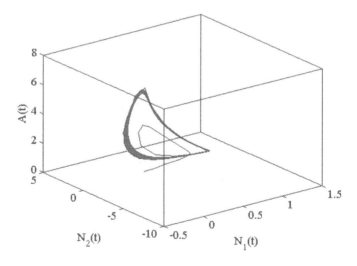

FIGURE 4.6 The phase diagram of the model system (4.1) with respect to the bifurcating parameter τ. The effect of parameter β has been clearly observed in the optimization of innovation diffusion modeling.

4.7 CONCLUSION

It is quiet common to achieve the oscillatory nature in the processes of diffusion of an innovation. Here, in the chapter, an economic group model with the delay in adoption of an innovation has been given. The system (4.1) has been verified for basic results, and justified the positivity and boundedness of the solutions. All the solutions of model (4.1) are lying inside the boundary of a defined region. The locally asymptotic stability behavior of model (4.1) is investigated for the adopter-free U^0 and nonnegative steady state U^*.

The system (4.1) established locally stable behavior without any evaluation period. But the delay parameter τ in the system (4.1) pushes the system into Hopf bifurcation, i.e., small amplitude periodic oscillations have been anticipated around the positive equilibrium U^*. This means that as and when the value of τ crossed over the threshold value $\tau_0 = 3.5$, the system is proven to be admitting stability switch and showed that there exists a limit cycle. More stable periodic oscillations of model (4.1) around U^* for non-adopter and adopter populations at $\tau = 4.3$ are shown in Figure 4.4. The bifurcation diagram of the system (4.1) helped to prove stable periodic orbits in the system, which has been exhibited in Figure 4.5. Also the role of internal factors (word of mouth interactions) has been discussed in optimization of innovation diffusion process of a high-valued product and proved that the excess of internal factors gave birth to periodic orbits with period doubling. It is evident that for the optimization process, the optimal use of internal factors in the form of word-of-mouth interactions will help the marketing managers to decide about the extent of factors and how they will diffuse a high-valued product in the social system.

REFERENCES

1. Bass, F. M., The Relationship between Diffusion Rates, Experience Curves, and Demand Elasticities for Consumer Durable Technological Innovations, Journal of Business, 53 (1980), pp. 551–567.
2. Bass, F. M., and Bultez, A. V., Optimal Strategic Pricing Policies with Learning, Krannert Graduate School of Management, Purdue Univ., Working Paper, (1979).
3. Dolan, R. J., and Jeuland, A. P., Experience Curves and Dynamic Demand Models: Implications for Optimal Pricing Strategies, Journal of Marketing, 45 (1981), pp. 52–62.
4. Dorfman, R., and Steiner, P. O., Optimal Advertising and Optimal Quality, American Economic Review, 44 (1954), pp. 826–836.
5. Erickson, G. M., Optimal Price-Advertising Interaction for New Consumer Durables, Univ. of Washington, Working Paper, (1982).
6. Gensch, D. H., and Whelam, U. P., An Optimum Budget Allocation Model for Dynamic, Interacting Market Segments, Management Science, 20 (1972), pp. 179–190.
7. Gensch, D. H., and Whelam, U. P., Optimal Price and Promotion for Interdependent Market Segments, Operations Research, 23 (1975), pp. 746–755.
8. Horsky, D., and Simon, L. S., Advertising in a Model of New Product Diffusion, in Proceedings of the first ORSA/TIMS Special Interest Conference on Market Measurement and Analysis, David B. Montgomery and Dick R. Wittink (eds.), Marketing Science Institute, Cambridge, MA, (1980), pp. 30–34.
9. Kalish, S., Monopolist Pricing with Dynamic Demand and Production Costs, Working Paper No. 8204, Graduate School of Management, University of Rochester, (1982).

10. Kalish, S., and Lilien, G. L., Optimal Subsidy Policy: Modeling the Federal Role in Accelerating the Diffusion of Innovation, Working Paper, Massachusetts Institute of Technology, (1980).

11. Kotowitz, Y., and Mathewson, F., Informative Advertising and Welfare, American Economic Review, 69 (1979), pp. 284–294.

12. Robinson, B., and Lakhani, C., Dynamic Price Models for New-Product Planning, Management Science, 21 (1975), pp. 1113–1122.

13. Sethi, S. P., Dynamic Optimal Control Models in Advertising: A Survey, SIAM Review, 19 (1977), pp. 685–725.

14. Teng, J. T., and Thompson, G. L., Oligopoly Models for Optimal Advertising When Production Costs Obey a Learning Curve, Management Science, 30 (1983), pp. 1087–1101.

15. Thompson, G. L., and Teng, J. T., Optimal Pricing and Advertising Policies for New Product Oligopoly Models, Marketing Science, 3 (1984), pp. 148–168.

16. Sethi, S. P., Optimal Advertising for the Nerlove-Arrow Model under a Budget Constraint, Operational Research Quarterly, 28 (1977), pp. 683–693.

17. Kumar, R., Sharma, A. K., and Agnihotri, K., Dynamical Analysis of an Innovation Diffusion Model with Evaluation Period, Boletim da Sociedade Paranaense de Matemática, 38(5) (2020), pp. 87–104.

18. Kumar, R., Sharma, A. K., and Agnihotri, K., Dynamics of an Innovation Diffusion Model with Time Delay, East Asian Journal on Applied Mathematics, 7(3) (2017), pp. 455–481.

19. Hale, J. K., Functional Differential Equations, Springer, Berlin, Heidelberg, (1971).

20. Aiello, W. G., and Freedman, H., A Time-Delay Model of Single-Species Growth with Stage Structure, Mathematical Biosciences, 10(2) (1990), pp. 139–153.

21. Birkhoff, G., and Rota, G., Ordinary Differential Equations, Ginn, Boston, MA, (1989).

22. Boonrangsiman, S., Bunwong, K., and Moore, E. J., A Bifurcation Path to Chaos in a Time-Delay Fisheries Predator–Prey Model with Prey Consumption by Immature and Mature Predators, Mathematics and Computers in Simulation, 124 (2016), pp. 16–29.

23. Sharma, A., Sharma, A. K., and Agnihotri, K., The Dynamic of Plankton–Nutrient Interaction with Delay, Applied Mathematics and Computation, 231 (2014), pp. 503–515.

24. Kumar, R., Sharma, A. K., and Agnihotri, K., Hopf Bifurcation Analysis in a Multiple Delayed Innovation Diffusion Model with Holling II Functional Response, Mathematical Methods in the Applied Sciences, 43(4) (2020), pp. 2056–2075.

5 TODIM-Based Multi-Criteria Decision-Making Approach for Selecting Inventory Policy

Anjali Priya
Department of Mathematics, Birla Institute
of Technology Mesra, Ranchi, India

Prabjot Kaur
Birla Institute of Technology Mesra, Ranchi, India

CONTENTS

5.1 ROLES OF INVENTORY

The primary role of an inventory system is to track your products and supplies. Although an effective inventory policy is a worthy theoretical goal, it is difficult to implement in practice. We all understand the importance of maintaining inventories to keep a business running. One of the management's responsibilities is to devise policies for the most efficient creation and distribution of inventories. Many factors

DOI: 10.1201/9781003156291-5

influence management's decision to adopt a policy. To a large extent, these policies represent the environment in which a business operates. The importance of inventories in a company's or supply chain's strategy is influenced by environmental factors.

5.2 FUNDAMENTAL QUESTIONS

When it comes to inventories, four basic questions must be answered. The first is what should be kept in a system's inventory? The response is dependent on a company's goals and the approach used to achieve those goals. Although Walmart and Flipkart are both retailers, they are fundamentally different. The goods they offer are one of the ways they differ. Thousands of various item kinds may be in stock at Walmart stores. Flipkart offers over 40 million different product categories from which one needs to choose. As a result, the breadth of a company's product offering is a critical choice.

The second concern is where the product should be stored. Should every store in a chain stock the same items? Flipkart has no physical sites. From each of its locations, the company services its own consumers and third-party warehouses. What should each warehouse have on hand? Should all things be stocked in every location, or should some goods be stocked only in one?

The final question is how much should be ordered when placing an order? Several things impact the answer to this question, which we'll discuss in the next section. The answer to the fourth question, "When should an order be placed?" will be determined by these considerations.

5.3 FACTORS INFLUENCING INVENTORY POLICY DECISIONS

Several key factors must be considered when developing mathematical models to respond to the issues stated in the previous section. Models are, by definition, abstractions or representations of real-world operating environments. As a result, not all the variables that influence the model capture or reflect inventory policy decisions. In this chapter, we will look at a variety of models, each with its way of expressing the individual factors mathematically. Let's talk about the underlying factors that influence inventory policy decisions before we dive into these mathematical models.

5.3.1 THE SYSTEM STRUCTURE

The supply chain's structure will be considered initially. In such a system, a central warehouse may carry a variety of item kinds sourced from several production suppliers. Then all the items are redistributed to regional centers, sometimes across the globe, according to some policy as shown in Figure 5.1.

5.3.2 THE ITEMS

A second issue to consider is the sort of commodities that are stocked in the supply chain and at a certain location. When it comes to determining stocking policies, the number of things in stock and how they interact are critical. Physical characteristics

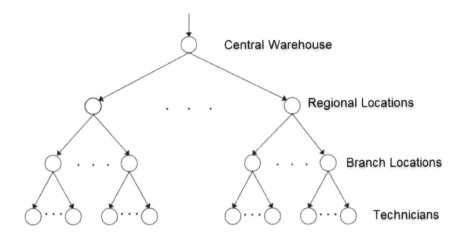

FIGURE 5.1 Supply chain example [31].

of objects vary. They are different in terms of weight and volume. At times, products may be perishable. Items such as hospital supplies, food, and chemicals are just a few examples with short shelf lives that must be carefully maintained.

In the eyes of the consumer, certain products are not special. Substitutions become increasingly common as the number of products of essentially the same kind (soft drinks, for example) grows. As a result, it is customary to substitute one product for another. The demands of the market vary from one item to the next. Demand rates and variability are not the same. A consumable product, like food, or a repairable product, such as a washing machine, is another important attribute of the products. Lastly, the most visible distinction between items is their price. An automobile engine is far more expensive than a toothpick. As a result, policies for regulating items will be heavily influenced by the cost of the items as well as the costs of storing and maintaining them. The nature of these expenditures will be explained in the following paragraphs.

5.3.3 LEAD TIMES

In case of supply chains, lead times exist, having a significant impact on stock levels. In general, it refers to the amount of time in between placing an order and receiving it. The longer it takes, the more unsure we are about the demand over time, and as a result, inventory is required. We consider the length of a supplier's lead time to be a measure of responsiveness.

5.3.4 COSTS

One of the most important considerations in deciding on an inventory policy is cost. Most inventory planning models consider a variety of costs. Obsolescence costs include costs such as purchasing, carrying, or holding, stockout, and other costs that are encountered in practice and recorded in models. When placing orders, the unit and fixed costs that are incurred will determine the inventory policies used to keep track of the materials, components, and completed goods that are purchased. The price per unit may vary depending on the quantity bought. The costs of procurement that are incurred regardless of the quantity purchased are known as fixed procurement costs.

There are also trade-offs to consider. Firms spend resources because of their frequent purchases. As the frequency of purchases increases, so does the number of fixed expenses incurred; yet, the cost of maintaining or transferring inventory for most commodities for short periods of time is very minimal. The opportunity cost of not investing capital in other ventures is included in a company's carrying costs; insurance, taxes, damages, pilferage, and warehouse operating costs (fixed and variable); out-of-pocket payments for insurance, taxes, damages, pilferage, and warehouse operating costs (fixed and variable); as well as the price of obsolescence. When supply exceeds demand, holding costs are incurred; nevertheless, there may not be enough inventory on hand to meet all demand at the time demand occurs.

5.4 MEASURING PERFORMANCE

Inventory management systems are accustomed to delivering goods and services to consumers. The quality service demanded is usually rather high, though it varies depending on the sort of product, the consumer being supplied, and the timeliness of both the need and the available or required supply. Supply chains are developed, and depending on the nature of the product and the needs of the customer, the physical and management infrastructures inherent in supply chain architecture are used to serve demand via inventory strategies. The inventory techniques that are used are mostly based on a model. These models can either be extremely simple or complex. The complexity of a supply chain is determined by the components that represent demand, prices, constraints, and the complexities of the supply chain. Regardless of the model's complexity, the implementation of the model must accomplish some objectives. When introduced, the aim can be to create a policy that either maximizes benefit or minimizes the expenditures that are relevant. Obviously, the right strategy would be dictated by assumptions about how expenditures are incurred, how requests arise and are supplied, and other factors.

In this chapter, TODIM-based multi-criteria decision-making (MCDM) approach is proposed for evaluating inventory policy under Pythagorean fuzzy environment. Ranking of inventory policy is a problem. Since the multi-criteria question involves unclear and conflicting qualitative and quantitative variables, such variables are not necessarily clear and often suffer vagueness owing to the inconsistent existence of the gathered data. In the Pythagorean fuzzy setting, they are therefore considered to eliminate this vagueness by assigning membership and non-member roles to these factors. Inventory policy selection and evaluation include a procedure that can be thought about, checked after some time, and ideally improved through certain measures. Determination of the right inventory control policy is a challenge in the dynamic business environment as it enables organizations to gain an upper hand in terms of cost, quality, and service, which in turn provides a vital step in fulfilling the requirements of the customer. This chapter illustrates the method to determine the best inventory control policy to be practiced based on major criteria such as raw material ordering frequency, stock verification, production type, inventory cost, capital investment, and demand policies by any large-, medium-, or small-scale firm to maximize their productivity and profit.

5.5 MULTI-CRITERIA DECISION-MAKING (MCDM)

The goal of MCDM is to select the best option from a finite number of options, which best meets several criteria. It is a decision science's most important component, and its principles have been widely utilized in domains such as economics [1–3], management [4–6], engineering [7–9], and others. TOPSIS (Technique for Order of Preference by Similarity to Ideal Solution) [10], ELECTRE (Elimination and Selection Expressing Reality) [11], and PROMETHEE [12] are some of the approaches presented to deal with MCDM difficulties. The complexity of real-world MCDM difficulties is growing, and risk attitudes of decision-makers (DMs) must eventually be considered during the MCDM process. Initiated by Kahneman and Tversky, the prospect theory is a concise framework for under-risk decision-making. This theory encompasses three essential aspects [13]:

1. Reliance on a source of information. According to an alternate method, the results are represented as gains and losses.
2. Sensitivity is being reduced. For profit, the DMs are risk cautious. They are, however, in high danger of losing money.
3. Failure is a fear. Losses are significantly more dangerous to the DMs than profits [14].

Based on the prospect theory, Gomes and Lima [15] first developed an MCDM method, called TODIM, which is true to solve MCDM problems where the psychological activities of the DMs are considered.

Furthermore, due to the complexity of the MCDM issues, it is difficult to provide extensive analyses of options in the DMs. To tackle this challenge, fuzzy set theory [16] was used for MCDM [17–19], which provides a vital way of explaining the complicated details. Later, intuitionist fuzzy set (IFS) [20], a fuzzy set extension, was found to be highly efficient in handling MCDM problems in the presence of uncertainty [21–23], using a degree of acceptance and a degree of nonacceptance, the sum of which is less than or equal to one, to assess a thing from both positive and negative perspectives at the same time. The Pythagorean fuzzy set (PFS) [33] is a useful tool for illustrating uncertainty in MCDM issues. The PFS [24, 25] has recently gained popularity as a technique for simulating unclear MCDM scenarios. The PFS is more extensive than the IFS, and it is often defined by the degree of membership and nonparticipation, with multiple squares equal to or less than 1. If a DM specifies a membership degree of 0.8 and a non-membership degree of 0.6, for example, the PFS will handle the issues that the IFS cannot, and it will then be valid only for the PFS.

To put it another way, all intuitive fuzzy degrees are included in the Pythagorean fuzzy degrees, meaning that the PFS is more successful in dealing with perplexing situations. Because of the nature of the events and the limitations of our human thinking, we are rarely able to make accurate conclusions. The PFS can express this type of ambiguity effectively through positive and negative assessments, which may successfully communicate the DMs' bewilderment and uncertainty in MCDM situations. Based on the PFS' enormous benefits, we plan to extend the TODIM technique to tackling MCDM difficulties utilizing Pythagorean fuzzy knowledge in this study.

Prior research [26–28] has taken pains to capture the DMs' attitudinal traits in MCDM situations, which are useful in expressing the DMs' attitudes toward unknown and uncertain objects. Uncertainty and risk are both frequent factors that must be considered in the MCDM cycle due to the magnitude of the MCDM challenges. As a result, we feel it is vital to evaluate the DMs' risk attitudes in the MCDM issue. TODIM, as previously mentioned, is an effective approach for capturing DMs' psychological behavior when they are under duress. Several scholars [29, 30] have utilized the fuzzy TODIM approach to concurrently depict the complexity and challenge of MCDM situations.

5.6 PYTHAGOREAN FUZZY SETS

PFSs were pioneered by Yager [8], to deal with vagueness of the membership grades as pairs satisfying the conditions of membership and non-membership degrees.

Let a set X be a universal set, then the PFS P is defined as the following:

$$P = \left\{ \begin{array}{l} \langle x, P(\mu_p(x), \nu_p(x)) \rangle \mid x \in X, \\ 0 \leq (\mu_p(x))^2 + (\nu_p(x))^2 \leq 1 \end{array} \right\} \tag{5.1}$$

where μ_p denotes the degree of membership, and ν_p denotes the degree of non-membership function of elements x to P.

$P(\mu_p(x), \nu_p(x))$ is expressed as $\beta = P(\mu_p, \nu_p)$. Also, the square sum of μ and ν is not more than one, i.e., $\mu_\beta^2 + \nu_\beta^2 \leq 1$, where $\mu_\beta, \nu_\beta \in [0,1]$.

Some of the operations defined on PFS are as below:

$$\beta_1 \oplus \beta_2 = P\left(\sqrt{\mu_{\beta1}^2 + \mu_{\beta2}^2 - \mu_{\beta1}^2 \nu_{\beta2}^2}, \nu_{\beta1} \nu_{\beta2}\right) \tag{5.2}$$

$$\beta_1 \otimes \beta_2 = P\left(\mu_{\beta1} \mu_{\beta2}, \sqrt{\nu_{\beta1}^2 + \nu_{\beta2}^2 - \nu_{\beta1}^2 \nu_{\beta2}^2}\right) \tag{5.3}$$

$$\lambda\beta = P\left(\sqrt{1 - (1 - \mu_\beta^2)^\lambda}, \nu_\beta^2\right), \lambda > 0 \tag{5.4}$$

$$\beta^\lambda = P\left((\mu_\beta)^\lambda, \sqrt{1 - (1 - \nu_\beta^2)^\lambda}\right), \lambda > 0 \tag{5.5}$$

5.7 THE TODIM APPROACH

The TODIM technique [15], which was proposed to understand the DM's psycho-logical actions, should be able to deal with the MCDM challenges satisfactorily. This methodology generates a multi-criteria value function based on the prospect principle to portray the superiority of each choice over others [13].

Let C_j $(j = 1, 2, 3, \ldots)$ be the referenced criterion, the weight vector w of which $= (w_1, w_2, \ldots, w_n)^T$ with $w_j \in [0,1]$ $(j = 1, 2, \ldots, n)$. Then the decision matrix $A = (a_{ij})_{n \times m}$ can be formed for alternatives Y_i $(i = 1, 2, \ldots, m)$, where a_{ij} is the DM's value for the criteria over the alternatives Y_i concerning the criterion C_j. We define

$$w_{jr} = \frac{w_j}{w_r} \tag{5.6}$$

as a relative weight of the criterion C_j to C_r and

$$w_r = \max\left\{w_j \mid j = 1, 2, \ldots, n\right\} \tag{5.7}$$.

The procedure is explained in the following section.

5.8 PYTHAGOREAN FUZZY TODIM

1. *Definition:* Suppose $\beta = P(\mu_\beta, \nu_\beta)$ is a PFN, then there's the scoring func-tion of β, which will be stated as [4]

$$s(\beta) = (\mu_\beta)^2 - (\nu_\beta)^2 \tag{5.8}$$

The better the PFN, the higher is the score.

We already know that the score function works well for comparing the vast majority of PFNs. If we have $\beta_1 = P(0.6, 0.6)$ and $\beta_2 = P(0.09, 09)$, then $s(\beta_1) = s(\beta_2) = 0$; however, we know that these PFNs are not the same.

2. *Definition:* Suppose $\beta = P(\mu_\beta, \nu_\beta)$ is a PFN, the accuracy of function β [33] will be defined as the following:

$$h(\beta) = (\mu_\beta)^2 + (\nu_\beta)^2 \tag{5.9}$$

where $0 \le h(\beta) \le 1$. The greater the value $h(\beta)$, the higher is the accuracy. We also express the hesitant degree of $x \in X$ as the following:

$$\pi_\beta(x) = \sqrt{1 - \mu_\beta^2(x) - \nu_\beta^2(x)} \tag{5.10}$$

Also,

$$\pi_\beta^2 + h(\beta) = 1 \tag{5.11}$$

As a result, the better the accuracy of PFNs, the lower is the hesitant degree.

3. *Definition:* Suppose $\beta_1 = P(\mu_{\beta 1}, v_{\beta 1})$ and $\beta_2 = P(\mu_{\beta 2}, v_{\beta 2})$ are two PFNs. Let $s(\beta_i)(i=1,2)$ and $h(\beta_i)(i=1,2)$ be the score values and accuracy values of β_1 and β_2, respectively [33], then

- If $s(\beta_1) < s(\beta_2)$, then $\beta_1 < \beta_2$;   (5.12)
- If $s(\beta_1) = s(\beta_2)$, then   (5.13)

If $h(\beta_1) < h(\beta_2)$, then $\beta_1 < \beta_2$

If $h(\beta_1) = h(\beta_2)$, then $\beta_1 = \beta_2$

If $h(\beta_1) > h(\beta_2)$, then $\beta_1 > \beta_2$

4. *Definition:* Suppose $\beta_1 = P(\mu_{\beta 1}, v_{\beta 1})$ and $\beta_2 = P(\mu_{\beta 2}, v_{\beta 2})$ are two PFNs. The Euclidean distance [32] between β_1 and β_2 becomes the following [33]:

$$d(\beta_1, \beta_2) = \sqrt{\frac{1}{2}\left[\begin{array}{c}\left((\mu_{\beta 1})^2 - (\mu_{\beta 2})^2\right)^2 + \left((v_{\beta 1})^2 - (v_{\beta 2})^2\right)^2 \\ + \left((\pi_{\beta 1})^2 - (\pi_{\beta 2})^2\right)^2\end{array}\right]} \qquad (5.14)$$

Definition: Let $\beta_1 = P(\mu_{\beta 1}, v_{\beta 1})$ and $\beta_2 = P(\mu_{\beta 2}, v_{\beta 2})$ be two PFNs, then

- $d(\beta_1, \beta_2) = d(\beta_2, \beta_1)$ (5.15)
- $d(\beta_1, \beta_2) = 0$ only if $\beta_1 = \beta_2$ (5.16)
- $0 < d(\beta_1, \beta_2) < 1$ (5.17)

5.9   PROCEDURE FOR PYTHAGOREAN FUZZY TODIM

Let A_i be the alternative and C_j be the criteria and PFN be expressed as $r_{ij} = P(\mu_{ij}, v_{ij})$. Then the Pythagorean fuzzy decision matrix is represented as $R = (r_{ij})_{m \times n}$,

$$R = \begin{bmatrix} P(\mu_{11}, v_{11}) & P(\mu_{12}, v_{12}) & \cdots & P(\mu_{1n}, v_{1n}) \\ P(\mu_{21}, v_{21}) & P(\mu_{22}, v_{22}) & \cdots & P(\mu_{2n}, v_{2n}) \\ \vdots & \vdots & \cdots & \vdots \\ P(\mu_{m1}, v_{m1}) & P(\mu_{m2}, v_{m2}) & \cdots & P(\mu_{mn}, v_{mn}) \end{bmatrix} \qquad (5.18)$$

where criteria are along the row and alternatives along the column. We follow the steps as follows:

Step 1: The DM provides the Pythagorean decision fuzzy matrix $R = (r_{ij})_{m \times n}$, which we create.

Step 2: By transforming this matrix into a normalized decision matrix $L = (l_{ij})_{m \times n}$, where

$$l_{ij} = \begin{cases} r_{ij} \\ (r_{ij})^c \end{cases} \qquad (5.19)$$

And r_{ij} is for benefit criteria and $(r_{ij})^c$ is for the cost criteria.

Step 3: We then calculate the relative weight of each criterion as $w_{jr} = w_j/w_r$, where w_j is the weight of criteria C_j and

$$w_r = \max\left\{w_j \mid j = 1, 2, \ldots, n\right\} \tag{5.20}.$$

Step 4: We calculate the degree of dominance [32] of each alternative A_i over A_j for criterion C_j by

$$\phi(A_i, A_t) = \begin{cases} \sqrt{w_{jr}\, d\left(l_{ij}, l_{tj}\right)\Big/ \sum_{j=1}^{n} w_{jr}} & l_{ij} > l_{tj} \\ 0 & l_{ij} = l_{tj} \\ -\dfrac{1}{\theta}\sqrt{\left(\sum_{j=1}^{n} w_{jr}\right) d\left(l_{ij}, l_{tj}\right)/w_{jr}} & l_{ij} < l_{tj} \end{cases} \tag{5.21}$$

where $d\left(l_{ij}, l_{tj}\right)$ is the distance between the PFN l_{ij} and l_{tj} and θ is the attenuation factor of the losses.

Using this above formula, we construct a dominance degree matrix to show $\varphi_j = \varphi_j\left(A_i, A_t\right)_{m \times m}$ visually for j=1, 2...n

$$
\varphi_j = \begin{array}{c} \\ A_1 \\ A_2 \\ \vdots \\ A_m \end{array}
\overset{\displaystyle A_1 A_2 \cdots A_m}{
\begin{bmatrix}
0 & \varphi_j(A_1, A_2) & \cdots & \varphi_j(A_1, A_m) \\
\varphi_j(A_2, A_1) & 0 & \cdots & \varphi_j(A_2, A_m) \\
\vdots & \vdots & 0 & \vdots \\
\varphi_j(A_m, A_1) & \varphi_j(A_m, A_2) & \cdots & 0
\end{bmatrix}} \tag{5.22}
$$

Step 5: The entire dominance degree is then calculated using the formula:

$$\delta(A_i, A_t) = \sum_{j=1}^{n} \varphi_j(A_i, A_t) \tag{5.23}$$

Step 6: We then derive the overall value [32] of each alternative A_i by

$$\xi_i = \frac{\sum_{t=1}^{m} \delta(A_i, A_t) - \min_i\left\{\sum_{t=1}^{m} \delta(A_i, A_t)\right\}}{\max_i\left\{\sum_{t=1}^{m} \delta(A_i, A_t)\right\} - \min_i\left\{\sum_{t=1}^{m} \delta(A_i, A_t)\right\}} \tag{5.24}$$

And the concept may be used to rank the order of each alternative: the higher the overall value ξ_i, the better is the alternative A_i.

Step 7: We provide a ranking to the options based on their overall values.

We can find the ranking of several inventory policies that will boost the enterprise performance by applying the Pythagorean fuzzy TODIM methods outlined above.

5.10   RESULT

The TODIM method, built based on the prospect principle, will effectively explain under-risk psychological actions of the DM, whereas the PFS is an efficient conceptual strategy for representing ambiguity. The Pythagorean fuzzy TODIM methodology is presented in this chapter, and it may be utilized to handle MCDM challenges involving the DMs' psychological behavior in ambiguous scenarios. The simulations were run to evaluate how the risk mindset influences the MCDM problem findings. The chapter has presented an appropriate guide to coping with the MCDM issues in this dynamic environment, which can not only depict the confusion but can also reflect the dangerous psychological actions of the DMs. In more analysis, the method would address the qualitative and quantitative details, which can also depict the ambiguity and the risk.

5.11   NOTATION

Symbols	Meaning
$P(x, y)$	Pythagorean set
μ_p	Degree of membership
v_p	Degree of non-membership
$\oplus$	Direct sum
$\otimes$	Kronecker product of two matrices
$s(\beta_i)$	Score values
$h(\beta_i)$	Accuracy values
$\varphi(A_i, A_t)$	Degree of dominance
ξ_i	Overall values of the alternatives

REFERENCES

1. A. Rabbani, M. Zamani, A. Yazdani-Chamzini, E. K. Zavadska, Proposing a new integrated model based on sustainability balanced scorecard (SBSC) and MCDM approaches by using linguistic variables for the performance evaluation of oil producing companies, *Expert Systems with Applications*, 41 (16) (2014) 7316–7327.
2. G. Kou, Y. Peng, C. Lu, MCDM approach to evaluating bank loan default models, *Technological and Economic Development of Economy*, 20 (2) (2014) 292–311.
3. E. Mulliner, N. Malys, V. Maliene, Comparative analysis of MCDM methods for the assessment of sustainable housing affordability, *Omega*, 59 (2015), 146–156.
4. M. Li, L. Jin, J. Wang, A new MCDM method combining QFD with TOPSIS for knowledge management system selection from the user's perspective in intuitionistic fuzzy environment, *Applied Soft Computing*, 21 (2014) 28–37.

5. E. Celik, A. T. Gumus, M. Alegoz, A trapezoidal type-2 fuzzy MCDM method to identify and evaluate critical success factors for humanitarian relief logistics management, *Journal of Intelligent & Fuzzy Systems: Applications in Engineering and Technology*, 27 (6) (2014) 2847–2855.
6. S. Sakhuja, V. Jain, F. Dweiri, Application of an integrated MCDM approach in selecting outsourcing strategies in hotel industry, *International Journal of Logistics Systems and Management*, 20 (3) (2015) 304–324.
7. Y. H. Hung, S. C. T. Chou, G. H. Tzeng, Knowledge management adoption and assessment for SMEs by a novel MCDM approach, *Decision Support Systems*, 51 (2011) 270–291.
8. T. Gürbüz, Y. E. Albayrak, An engineering approach to human resources performance evaluation: Hybrid MCDM application with interactions, *Applied Soft Computing*, 21 (2014) 365–375.
9. S. Hashemkhani Zolfani, R. Maknoon, E. K. Zavadskas, Multiple nash equilibriums and evaluation of strategies. New application of MCDM methods, *Journal of Business Economics and Management*, 16 (2) (2015) 290–306.
10. C. L. Hwang, K. S. Yoon, *Multiple Attribute Decision Methods and Applications*, Berlin: Springer, 1981.
11. B. Roy, P. Bertier, *La méthode ELECTRE II: uneméthode de classementenprédence de critères multiples*, 1971.
12. B. Mareschal, J. P. Brans, P. Vincke, *PROMETHEE: A New Family of Outranking Methods in Multicriteria Analysis*, ULB, Universite Libre de Bruxelles, 1984.
13. D. Kahneman, A. Tversky, Prospect theory: An analysis of decision under risk, *Econometrica: Journal of the Econometric Society*, 47 (1979) 363–391.
14. M. Abdellaoui, H. Bleichrodt, C. Paraschiv, Loss aversion under prospect theory: A parameter-free measurement, *Management Science*, 53 (2007) 1659–1674.
15. L. Gomes, M. Lima, TODIM: Basics and application to multicriteria ranking of projects with environmental impacts, *Foundations of Computing and Decision Sciences*, 16 (1992) 113–127.
16. L. A. Zadeh, Fuzzy sets, *Information Control*, 8 (1965) 338–353.
17. R. E. Bellman, L. A. Zadeh, Decision-making in a fuzzy environment, *Management Science*, 17 (1970) 141–161.
18. R. R. Yager, Fuzzy decision making including unequal objectives, *Fuzzy Sets and Systems*, 1 (1978) 87–95.
19. K. Nakamura, Preference relations on a set of fuzzy utilities as a basis for decision making, *Fuzzy Sets and Systems*, 20 (1986) 147–162.
20. K. T. Atanassov, Intuitionistic fuzzy sets, *Fuzzy Sets and Systems*, 20 (1986) 87–96.
21. Z. S. Xu, R. R. Yager, Dynamic intuitionistic fuzzy multi-attribute decision making, *International Journal of Approximate Reasoning*, 48 (2008) 246–262.
22. H. W. Liu, G. J. Wang, Multi-criteria decision-making methods based on intuitionistic fuzzy sets, *European Journal of Operational Research*, 179 (2007) 220–233.
23. F. E. Boran, S. Genç, M. Kurt, et al., A multi-criteria intuitionistic fuzzy group decision making for supplier selection with TOPSIS method, *Expert Systems with Applications*, 36 (2009) 11363–11368.
24. R. R. Yager, Pythagorean fuzzy subsets, in *Proceeding of The Joint IFSA World Congress and NAFIPS Annual Meeting*, Edmonton Canada, pp. 57–61 (2013).
25. R. R. Yager, Pythagorean membership grades in multicriteria decision making, *IEEE Transactions on Fuzzy Systems*, 22 (2014) 958–965.
26. L. H. Chen, C. C. Hung, C. C. Tu, Considering the decision maker's attitudinal character to solve multi-criteria decision-making problems in an intuitionistic fuzzy environment, *Knowledge-Based Systems*, 361 (2012) 29–38.
27. H. C. Liu, J. X. You, X. J. Fan, et al., Site selection in waste management by the VIKOR method using linguistic assessment, *Applied Soft Computing*, 214 (2014) 53–61.

28. J. Wu, F. Chiclana, A risk attitudinal ranking method for interval-valued intuitionistic fuzzy numbers based on novel attitudinal expected score and accuracy functions, *Applied Soft Computing*, 22 (2014) 272–286.

29. R. A. Krohling, T. T. M. de Souza, combining prospect theory and fuzzy numbers to multi-criteria decision making, *Expert Systems with Applications*, 39 (2012) 11487–11493.

30. Z. P. Fan, X. Zhang, F. D. Chen, et al., Extended TODIM method for hybrid multiple attribute decision making problems, *Knowledge-Based Systems*, 42 (2013) 40–48.

31. J. A. Muckstadt, *Inventories Are Everywhere*, Springer Series in Operations Research and Financial Engineering, 2010.

32. Y. Li, Y. Shan, P. Liu, An extended TODIM method for group decision making with the interval intuitionistic fuzzy sets, 2015, 1–9. Article ID 672140.

33. P. J. Ren, Z. S. Xu, X. J. Gou, Pythagorean fuzzy TODIM approach to multi-criteria decision making, *Applied Soft Computing*, 42 (2016) 246–259.

6 Performability and Sensitivity of a System Equipped with *k*-Out-of-*n* Redundancy

Pardeep Kumar
Faculty of Chemical Engineering and Physical Sciences,
Lovely Professional University, Phagwara, Punjab

Amit Kumar
Department of Mathematics, Symbiosis Institute of Technology,
Symbiosis International University, Pune, India

Mangey Ram
Department of Mathematics, Computer Science & Engineering,
Graphic Era Deemed to be University,
Dehradun, Uttarakhand

CONTENTS

DOI: 10.1201/9781003156291-6

97

6.1   INTRODUCTION

The complexity in the industrial systems is increasing day by day with the rapid development in industry. A system is a combination of many components. System's reliability depends on the reliability of its components. The whole system may fail when a different component of the system fails. Therefore, the improvement in system reliability is necessary. For improving system reliability and availability, redundancy technique is used. Many useful work have been done in the past regarding the uses of redundancy. Some authors also paid their attention to *k-out-of-n: F* redundancy. El-Damcese and El-Sodany [1] discussed a *k-out-of-n: F* redundancy in a system and found reliability and sensitivity. Authors also did the comparative study of reliability measures concerning two dissimilar structures. Wang and Loman [2] focused on the reliability/availability of *k-out-of-n* system and incorporated "M" cold standby unit. A complex structure in series configuration was analyzed by Ram and Kumar [3]. This system comprises two substructures, "A" and "B," handled by human operator where substructure "A" is *2-out-of-3: F*, which is connected with the substructure "B" in series configuration. A system containing three indistinguishable units in parallel, which may work in two different states, operational and nonoperational, was analyzed by incorporating three types of failure, including partial failure, human failure and catastrophic failure by Ram and Manglik [4]. Bai et al. [5] determined the optimal number of redundant units in *k-out-of-n* system with common cause failure. Ardakan and Hamadani [6] used concept of mixed redundancy. With mixed redundancy policy, authors mean a system that is a combination of active and standby redundancies. In this system, every system can have different number of cold standby and active redundancies characterized by n_{A_i} and n_{S_i}. Authors determined values of n_{A_i} and n_{S_i} to optimize the reliability of system. Yusuf and Hussaini [7] analyzed the effects of repair action for a *2*-out-of-*3* system. Ding et al. [8] analyzed a framework for reliability approximation of multi-state weighted *k-out-of-n* system. Taghipour and Kassaei [9] analyzed load-sharing system. In the *k-out-of-n* system, *n* identical components share a certain amount of load. On the failure of any one component, its load is equally divided to the left behind *n−1* components. The author developed a model to find the optimal inspection interval for such a system, which reduces the total expected cost incurred over the system lifetime. Wu et al. [10] analyzed a *k-out-of-n: G* repairable system with a single repairman who takes a single vacation. In real-world situation, the repairman can perform other jobs in his idle time. The author determined how long should be the vacation time without affecting repairman's primary work. He et al. [11] investigated a repairable *k*-out-of-*n: G* system with repairman's multiple vacations. When there is no failure in components, the repairman goes for vacation. When he returns from his/her vacation, he/she goes for vacation again if there are less than *N* components waiting for repair. This happens until at least *N* broken components are waiting for repair. Krishnamoorthy and Ushakumari [12] investigated a *k-out-of-n: G* system with *D*-policy for the repair. Under this policy, when the workload surpasses a threshold limit *D*, a server is instantly called for repair and starts repair without wasting time. Grida et al. [13] compared the availability of a *3-out-of-4*: cold standby system and *6-out-of-8* system. Haggag [14] applied preventive maintenance incorporating for types of failure for a *3-out-of-4*: system. Kolmogorov's forward equations were obtained for a *k-out-of-n* repairable system with standby redundancy by Haggag [15]. Author proved that repair and standby systems play very important rule in the system. Mishra and Jainb [16] found for a main *k-out-of-n: F* secondary subsystem's steady state availability. The main subsystem shut off the secondary subsystem when more than *K* units of main system fail. A *M-out-of-N* warm standby system with "*R*" repair facilities was studied by Arulmozhi [17]. An algorithmic approach was presented by Barron et al. [18] for *R-out-of-n* system with several repairmen. Krishnan and Somasundaram [19] studied *k-out-of-n* system with sensor. Choudhury [20] analyzed Poisson queue with general step time and vacation period using supplementary variable technique.

6.2 DESCRIPTION OF THE SYSTEM

A system which consists of three subsystems, A, B and C, connected in series configuration, is analyzed in the present chapter. The nature of subsystem C is *2-out-of-3: F* with distinct components. The system can move in failed/degraded state due to unit or complete failure of any of its subsystem. The flow diagram of the system is shown in Figure 6.1.

6.3 ASSUMPTIONS

The following assumptions are taken into consideration throughout the analysis of the considered system

Assumption 1: The working stints of each unit of the system are supposed to be independent, i.e., components work independently.
Assumption 2: Maintenance team comes into action when any of the working components break down.
Assumption 3: Various processes (Markov) involved in the system are taken to be independent from each other.
Assumption 3: The system goes into failed/degraded state as soon as the unit's failure/failures occur. In this research shut-off rule is the suspended animation.
Assumption 4: Average failures rate are taken as constant.
Assumption 5: In a degraded state, system doesn't stop working.

6.4 NOMENCLATURE

Throughout the chapter, the following nomenclatures are followed. Table 6.1 gives a brief idea about the nomenclatures used in this chapter.

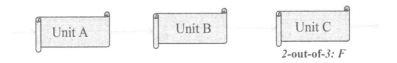

FIGURE 6.1   Configuration of the system.

TABLE 6.1
Nomenclature

t	*Time scale*
s	*Laplace Transformation variable*
$P_i(t);\ i = 0,1,2,3$	*Probability of the system being in state S_i at instant t.*
$\bar{P}_i(s)$	*Laplace transform of $P_i(t)$*
$P_j(x,t);$ $j = 4,5,6,7,8$	*Probability density function of system being in completely failed state at instant t with elapsed repair time x*
$\bar{P}_j(x,s)$	*Laplace transform of $P_j(x,t)$*
$\lambda_1/\lambda_2/\lambda_3$	*Failure rate of first/second/third component of the substructure C*
λ_A/λ_B	*Failure rate of substructure A and B*
$\mu_1/\mu_2/\mu_3$	*Represents repair rate of first/second/third component of substructure C*
μ_A/μ_B	*Repair rate of substructure A and B*
μ	*Simultaneous repair rate of any two units of subsystem C*
t	*Time scale*
s	*Laplace transformation variable*
$P_i(t);\ i = 0,1,2,3$	*Probability of the system being in state S_i at instant t*
$\bar{P}_i(s)$	*Laplace transform of $P_i(t)$*
$P_j(x,t);$ $j = 4,5,6,7,8$	*Probability density function of system being in completely failed state at instant t with elapsed repair time x*
$\bar{P}_j(x,s)$	*Laplace transform of $P_j(x,t)$*
$\lambda_1/\lambda_2/\lambda_3$	*Failure rate of first/second/third component of the substructure C*
λ_A/λ_B	*Failure rate of substructure A and B*
$\mu_1/\mu_2/\mu_3$	*Represents repair rate of first/second/third component of substructure C*
μ_A/μ_B	*Repair rate of substructure A and B*
μ	*Simultaneous repair rate of any two units of subsystem C*

6.5   STATE NARRATIVES

For various states, we follow the following narratives given in Table 6.2.

6.6   TRANSITION STATE DIAGRAM

By carefully analyzing the effects of various failures on the considered system, a transition state diagram is formulated (given in Figure 6.2), which successfully depicts the various stages of it throughout the functioning.

6.7   FORMULATION OF INTRO-DIFFERENTIAL EQUATIONS

The following set of intro-differential equations are generated with the help of Markov process and the above state transition diagram, which governed the present mathematical model.

TABLE 6.2
State Narratives

S_0	Good state: All the components of the system are in good working condition
S_1	Degraded state: State in which unit C is degraded due to failure of its first subunit
S_2	Degraded state: State in which unit C is degraded due to failure of its second subunit
S_3	Degraded state: State in which unit C is degraded due to failure of its third subunit
S_4	Failed state: State in which unit C is completely failed due to its first and third subunit failure
S_5	Failed state: State in which unit C is completely failed due to its first and second subunit failure
S_6	Failed state: State in which unit C is completely failed due to its second and third subunit failure
S_7	Failed state: State in which unit A is completely failed
S_8	Failed state: State in which unit B is completely failed

$$\left[\frac{\partial}{\partial t}+\lambda_1+\lambda_2+\lambda_3+\lambda_A+\lambda_B\right]P_0(t)=\mu_1 P_1(t)+\mu_2 P_2(t)+\mu_3 P_3(t)$$

$$+\int_0^\infty \mu_A P_7(x,t)\,dx+\int_0^\infty \mu_B P_8(x,t)\,dx+\sum_{i=4}^{6}\int_0^\infty \mu P_i(x,t) \qquad (6.1)$$

$$\left[\frac{\partial}{\partial t}+\lambda_3+\lambda_2+\mu_1\right]P_1(t)=\lambda_1 P_0(t) \qquad (6.2)$$

$$\left[\frac{\partial}{\partial t}+\lambda_1+\lambda_3+\mu_2\right]P_2(t)=\lambda_2 P_0(t) \qquad (6.3)$$

$$\left[\frac{\partial}{\partial t}+\lambda_2+\lambda_1+\mu_3\right]P_3(t)=\lambda_3 P_0(t) \qquad (6.4)$$

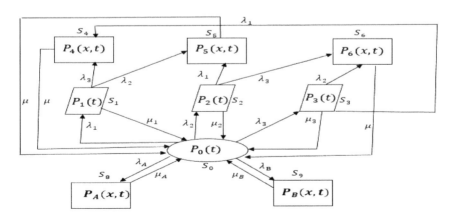

FIGURE 6.2 Transition state diagram.

$$\left[\frac{\partial}{\partial t}+\frac{\partial}{\partial x}+\mu\right]P_4(x,t)=0 \tag{6.5}$$

$$\left[\frac{\partial}{\partial t}+\frac{\partial}{\partial x}+\mu\right]P_5(x,t)=0 \tag{6.6}$$

$$\left[\frac{\partial}{\partial t}+\frac{\partial}{\partial x}+\mu\right]P_6(x,t)=0 \tag{6.7}$$

$$\left[\frac{\partial}{\partial t}+\frac{\partial}{\partial x}+\mu_A\right]P_7(x,t)=0 \tag{6.8}$$

$$\left[\frac{\partial}{\partial t}+\frac{\partial}{\partial x}+\mu_B\right]P_8(x,t)=0 \tag{6.9}$$

Boundary conditions

$$P_4(0,t)=\lambda_3\,P_1(t)+\lambda_1\,P_3(t) \tag{6.10}$$

$$P_5(0,t)=\lambda_1 P_2(t)+\lambda_2 P_1(t) \tag{6.11}$$

$$P_6(0,t)=\lambda_3 P_2(t)+\lambda_2 P_3(t) \tag{6.12}$$

$$P_7(0,t)=\lambda_A P_0(t) \tag{6.13}$$

$$P_8(0,t)=\lambda_B P_0(t) \tag{6.14}$$

Initial condition

$$P_i(t)=\begin{cases}1,\ t=0\ \text{and}\ i=0\\0,\ \ \text{otherwise}\end{cases} \tag{6.15}$$

The system of intro-differential (Equations 6.1–6.9) together with boundary conditions (Equations 6.9–6.14) and initial condition (Equation 6.15) is known as Chapman-Kolmogorov differential equations. In order to find the various performance indicator of the considered system, here authors solve the above set of equations with the help of Laplace transform and find various state probabilities $\overline{P}_i(s); i=0,1,2,3$ for the considered system as the following:

$$\overline{P}_0(s)=\frac{1}{T(s)} \tag{6.16}$$

$$\overline{P}_1(s)=A(s)\,\overline{P}_0(s) \tag{6.17}$$

$$\overline{P}_2(s)=B(s)\,\overline{P}_0(s) \tag{6.18}$$

$$\overline{P_3}(s)= C(s)\,\overline{P_0}(s) \tag{6.19}$$

where

$$T(s)=\begin{bmatrix}\left[s+\lambda_1+\lambda_2+\lambda_3+\lambda_A+\lambda_B\right]-\mu_1 A(s)-\mu_2 B(s)-\mu_3 C(s)-\dfrac{\mu_A\lambda_A}{s+\mu_A}-\dfrac{\mu_B\lambda_B}{s+\mu_B}\\[2mm]-\mu\left[\dfrac{\lambda_3 A(s)+\lambda_1 C(s)}{s+\mu}\right]-\mu\left[\dfrac{\lambda_1 B(s)+\lambda_2 A(s)}{s+\mu}\right]-\mu\left[\dfrac{\lambda_3 B(s)+\lambda_2 C(s)}{s+\mu}\right]\end{bmatrix}$$

$$A(s)=\frac{\lambda_1}{s+\lambda_3+\lambda_2+\mu_1}\ ,\quad B(s)=\frac{\lambda_2}{s+\lambda_3+\lambda_1+\mu_2}\ ,\quad C(s)=\frac{\lambda_3}{s+\lambda_3+\lambda_1+\mu_3}$$

The state probability for the various states of the considered system can be obtained by the aid of inverse Laplace transform of Equations (6.16)–(6.19). The numerical computation is carried out by taking various failure and repair as constant. The availability of the system has been computed as

$$A(t)=\sum_{i=0}^{3} P_i(t) \tag{6.20}$$

6.8   CALCULATION OF DIFFERENT PERFORMANCE INDICATORS

6.8.1   Availability

It is the probability that a system or equipment is available for operation when used under stated conditions in an ideal support environment. For evaluating the availability of the considered system, we put the value of various failures as $\lambda_1 = 0.03$, $\lambda_2 = 0.12$, $\lambda_3 = 0.08$, $\lambda_A = 0.3$, $\lambda_B = 0.25$ and repair as $\mu_1 = \mu_2 = \mu_3 = \mu = \mu_A = \mu_B = 1$ in Equation (6.20). The system availability is obtained as

$$A(t)=\begin{cases}0.006671e^{-1.15t}+0.0190283\,e^{-1.11t}+0.009287\,e^{-1.2t}\\[2mm]+8.470421\times10^{-13}\,e^{-0.89t}\left(\dfrac{4.094911\times10^{11}\sinh(0.861452t)+1.139272\times10^{12}}{\cosh(0.861452t)}\right)\end{cases}$$

$$\tag{6.21}$$

Now varying time unit t in Equation (6.21), we get Table 6.3 and corresponding Figure 6.3 for the availability of the considered system.

6.8.2   Reliability

Reliability is the probability of a system performing its purpose adequately for the intended period under the given operating conditions. The reliability of the

---

TABLE 6.3

Availability of the System

Time Unit (t)	Availability $A(t)$
0	1.00000
1	0.70229
2	0.63242
3	0.60486
4	0.58580
5	0.56885
6	0.55272
7	0.53714
8	0.52201
9	0.50732
10	0.49304

---

considered system is calculated by putting the value of various failure rates as $\lambda_1 = 0.03, \lambda_2 = 0.12, \lambda_3 = 0.08, \lambda_A = 0.3, \lambda_B = 0.25$ and all repair rates as zero in Equation (6.20); the reliability of the considered system is obtained as

$$R(t) = 0.051724 \ e^{-2t} + 0.047619 \ e^{-0.15t} + 0.721552 \ e^{-.78t} + 0.179104 \ e^{-.11t} \qquad (6.22)$$

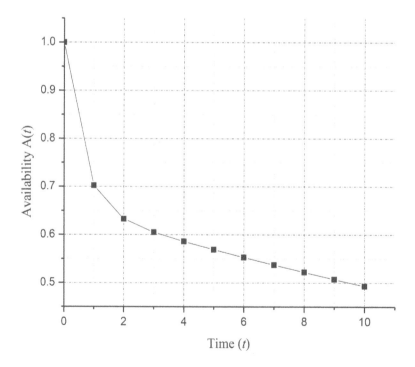

FIGURE 6.3 Behavior of availability of the system with time.

TABLE 6.4
Reliability of System

Time Unit (t)	Reliability R(t)
0	1.00000
1	0.57455
2	0.36531
3	0.25702
4	0.19659
5	0.15946
6	0.1342
7	0.11542
8	0.10048
9	0.08809
10	0.07754

Now varying time unit t in Equation (6.22), we get Table 6.4 and corresponding Figure 6.4 for the reliability of the considered system.

6.8.3 MEAN TIME TO FAILURE CALCULATION

Mathematically, the mean time to failure (MTTF) of a system is calculated as

$$MTTF = \int_0^t R(t)dt = \lim_{s \to 0} \bar{R}(s) \qquad (6.23)$$

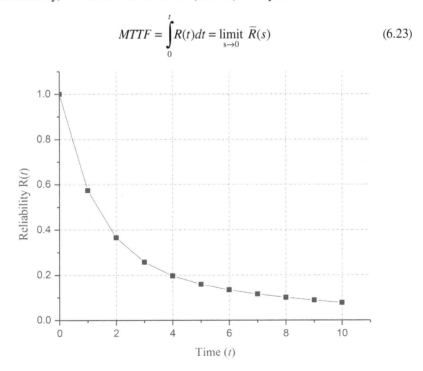

FIGURE 6.4 Behavior of reliability with time unit.

TABLE 6.5
MTTF of the System

Variations in $\lambda_1, \lambda_2, \lambda_3, \lambda_A, \lambda_B$	MTTF with Respect to Failure Rates				
	λ_1	λ_2	λ_3	λ_A	λ_B
0.1	3.083778966	3.034903692	2.824300699	4.208463950	3.874458875
0.2	3.214285714	3.553208672	2.017210146	3.589572193	3.343711084
0.3	3.361976370	4.059509571	1.635064935	3.129370629	2.940854326
0.4	3.494983277	4.498708565	1.397056432	2.773760331	2.624633431
0.5	3.610678532	4.873950394	1.229001826	2.490723562	2.369814651
0.6	3.710965868	5.195462023	1.101648352	2.260101010	2.160096541
0.7	3.798279097	5.472936612	1.000692091	2.068567026	1.984478936
0.8	3.874792808	5.714358055	.9181246727	1.906960227	1.835269993
0.9	3.942304195	5.926082366	.8490275144	1.768774704	1.706929434

Now using Equation (6.21) in Equation (6.23), we obtained the MTTF of the considered system as the following:

$$MTTF = \frac{\lambda_1^3 + 3\lambda_1^2\lambda_2 + 3\lambda_3\lambda_1^2 + 2\lambda_1\lambda_2^2 + 5\lambda_1\lambda_2\lambda_3 + 2\lambda_1\lambda_3^2 + \lambda_2^3 + 2\lambda_3\lambda_2^2 + \lambda_3^2\lambda_2}{(\lambda_3 + \lambda_2)(\lambda_1 + \lambda_3)(\lambda_1 + \lambda_2)(\lambda_1 + \lambda_2 + \lambda_3 + \lambda_A + \lambda_B)} \quad (6.24)$$

Varying failure rates one by one from 0.01 to 0.09 with an interval of 0.01, and in Equation 6.24, the MTTF of the considered system is obtained and shown in Table 6.5 and Figure 6.5.

6.8.4 Sensitivity Analysis for System's MTTF

It is a procedure that is used to decide how the different values of an independent failure influence a specific dependent parameter under some limitations or, in other way, it can be used as an indicator by which one can inspect that how the failures effect the system's performance. Here, the authors performed the sensitivity analysis of system's MTTF for finding how the failures affect the system MTTF.

Sensitivity analysis of the considered system with respect to MTTF is performed by differentiating the MTTF expression obtained in Equation (6.24) with respect to various failure rates and then placed the values of various failure rates as $\lambda_1 = 0.03, \lambda_2 = 0.12, \lambda_3 = 0.08$, $\lambda_A = 0.3$, $\lambda_B = 0.25$ in these partial derivatives. Now varying the failure rates one by one, respectively, in these partial derivatives, one can obtain Table 6.6 and corresponding Figure 6.6 for the sensitivity of MTTF for the system.

6.8.5 Expected Profit of the System

The expected profit of the system in the interval $[0, t]$ is given by

$$E_P(t) = K_1 \int_0^t A(t)dt - tK_2 \quad (6.25)$$

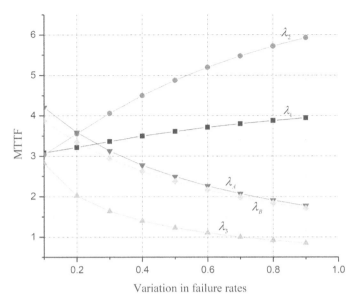

FIGURE 6.5   MTTF vs. variation in failure rates.

System-expected profit will be obtained by using Equation (6.20) in Equation (6.25). It is obtained as

$$E_P(t) = K_1 \left\{ \begin{array}{l} 23.183901 + 22.976747\,\mathrm{Sinh}(0.028547\,t) + 0.176469\,\mathrm{Sinh}(1.751452\,t) \\ -22.976747\,\mathrm{Cosh}(0.028547\,t) - 0.176469\,\mathrm{Cosh}(1.751452\,t) - 0.017142\ e^{-1.11t} \\ -0.005801\,e^{-1.15t} - 0.007739\,e^{-1.2t} \end{array} \right\} - K_2 t$$

$$(6.26)$$

TABLE 6.6
Sensitivity of the MTTF

Variation in Failure Rates $\lambda_1, \lambda_2, \lambda_3,$ λ_A, λ_B	System Sensitivity with Respect to MTTF				
	$\dfrac{\partial(MTTF)}{\partial\lambda_1}$	$\dfrac{\partial(MTTF)}{\partial\lambda_2}$	$\dfrac{\partial(MTTF)}{\partial\lambda_3}$	$\dfrac{\partial(MTTF)}{\partial\lambda_A}$	$\dfrac{\partial(MTTF)}{\partial\lambda_B}$
0.1	0.813948	0.228090	−13.18090	−7.25597	−6.14993
0.2	1.502081	−14.80231	−5.087348	−5.27878	−4.58042
0.3	1.416448	−24.44426	−2.907061	−4.01201	−3.54319
0.4	1.241713	−31.42442	−1.960911	−3.15200	−2.82218
0.5	1.075822	−36.80356	−1.445203	−1.28172	−2.30079
0.6	0.934074	−41.11105	−1.124509	−2.09268	−1.91158
0.7	0.815825	−44.65335	−0.907493	−1.75302	−1.61339
0.8	0.717435	−47.62494	−0.751839	−1.48981	−1.37990
0.9	0.635204	−50.15721	−0.635379	−1.28172	−1.19365

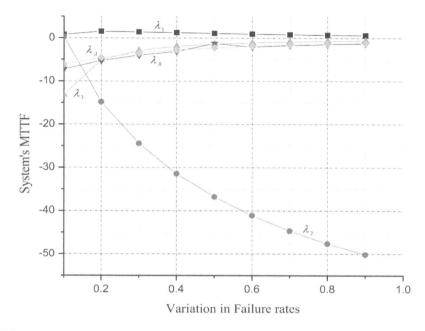

FIGURE 6.6 System's MTTF vs. variation in failure rates.

Now let $K_1 = 1$ and varying K_2 as 0.1, 0.2, 0.3, 0.4 and 0.5, respectively, in Equation (6.26) then we obtain time unit t as shown in Table 6.7 and corresponding Figure 6.7 for behavior of expected profit with time unit.

TABLE 6.7
Expected Profit of the System

Time Unit (t)	$K_2 = 0.1$	$K_2 = 0.2$	$K_2 = 0.3$	$K_2 = 0.4$	$K_2 = 0.5$
			Expected Profit from the System		
0	0	0	0	0	0
1	0.713374	0.613374	0.513374	0.413374	0.313374
2	1.273814	1.073814	0.873814	0.673814	0.473814
3	1.791115	1.491115	1.191115	0.891115	0.591115
4	2.286145	1.886145	1.486145	1.086145	0.686145
5	2.763375	2.263375	1.763375	1.263375	0.763375
6	3.224108	2.624108	2.024108	1.424108	0.824108
7	3.668989	2.968989	2.268989	1.568989	0.868989
8	4.098496	3.298496	2.498496	1.698496	0.898496
9	4.513198	3.613198	2.713198	1.813198	0.913198
10	4.913999	3.913999	2.913999	1.913999	0.913999

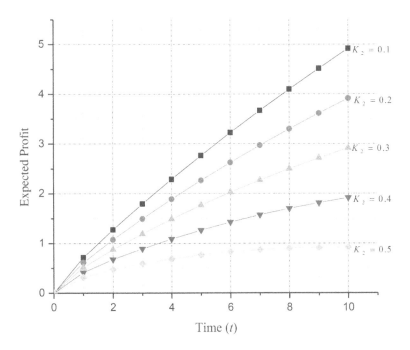

FIGURE 6.7   Expected profit vs. time unit and service cost.

6.9   RESULT DISCUSSION

One can say that the performance of a system directly/indirectly depends on the performance of each of its subsystems. So to increase the performance of the system, more attention should be given to their corresponding subunits for the effectiveness of the maintenance program. Here, an investigation is carried out for a system, which can be easily found in some of the industrial system, in order to find its most critical component, which affects the performance of the system. The effects of various failure rates on system availability, reliability and MTTF are analyzed and depicted with graphs. The graph of availability (Figure 6.3) and reliability (Figure 6.4) of the system shows that these decrease with respect to time in a smooth manner. It is also observed that the system availability after 10 units of time is 0.49304, while the reliability of the system after 10 units of time is 0.07754. The difference between these two shows the importance of maintenance policy. Figure 6.5 reflects the behavior of system's MTTF with respect to failure rates. Its shows that the system's MTTF is the highest with respect to the failure rate of the second unit of subsystem C and the lowest with respect to the third unit of subsystem C. Also, it is observed that system MTTF is approximately the same for the failure rate of subunit A and B. Sensitivity analysis for system's MTTF has been done and reflected in Figure 6.6. It shows that the system's MTTF is most sensitive with respect to the failure rate of the second unit of subsystem C and less sensitive for the failure rate of the first unit of subsystem C.

Equation (6.26) displays the expected profit from the system as a function of time and corresponding Figure 6.7 shows that the behavior of the profit functions with system service cost and time unit. One can say that expected profit from the system decreases as service cost and time unit increase.

6.10   CONCLUSION

On the basis of Section 6.8 (result discussion), one can say that the system reliability/availability can be controlled by controlling its various failure rates. It is also concluded the system performance specially MTTF is much more effected by the failure of the second unit/third unit of subsystem C. So, if one controls these failures, system performance can increase significantly. In future, this work can be extended by using some of the redundancy for subsystem C.

REFERENCES

1. El-Damcese, M. A., El-Sodany, N. H. (2015). Reliability and sensitivity analysis of the *k*-out-of-*n*: G warm standby parallel repairable system with replacement at common-cause failure using Markov model. *Reliability: Theory & Applications*, *10*(39), 4.
2. Wang, W., Loman, J. (2002). Reliability/availability of *k*-out-of-*n* system with M cold standby units. In *Reliability and Maintainability Symposium, 2002. Proceedings. Annual* (pp. 450–455). IEEE.
3. Ram, M., Kumar, A. (2014). Performance of a structure consisting a 2-out-of-3: F substructure under human failure. *Arabian Journal for Science and Engineering*, *39*(11), 8383–8394.
4. Ram, M., Manglik, M. (2014). Stochastic behaviour analysis of a Markov model under multi-state failures. *International Journal of System Assurance Engineering and Management*, *5*(4), 686–699.
5. Bai, D. S., Yun, W. Y., Chung, S. W. (1991). Redundancy optimization of *k*-out-of-*n* systems with common-cause failures. *IEEE Transactions on Reliability*, *40*(1), 56–59.
6. Ardakan, M. A., Hamadani, A. Z. (2014). Reliability optimization of series–parallel systems with mixed redundancy strategy in subsystems. *Reliability Engineering & System Safety*, *130*, 132–139.
7. Yusuf, I., Hussaini, N. (2012). Evaluation of reliability and availability characteristics of 2-out of-3 standby system under a perfect repair condition. *American Journal of Mathematics and Statistics*, *2*(5), 114–119.
8. Ding, Y., Zuo, M. J., Lisnianski, A., Li, W. (2010). A framework for reliability approximation of multi-state weighted k-out-of-n systems. *IEEE Transactions on Reliability*, *59*(2), 297–308.
9. Taghipour, S., Kassaei, M. L. (2015). Periodic inspection optimization of a k-out-of-n load-sharing system. *IEEE Transactions on Reliability*, *64*(3), 1116–1127.
10. Wu, W., Tang, Y., Yu, M., Jiang, Y. (2014). Reliability analysis of a *k*-out-of-*n*: G repairable system with single vacation. *Applied Mathematical Modelling*, *38*(24), 6075–6097.
11. He, G., Wu, W., Zhang, Y. (2018). Analysis of a multi-component system with failure dependency, N-policy and vacations. *Operations Research Perspectives*, *5*, 191–198.
12. Krishnamoorthy, A., Ushakumari, P. V. (2001). *K*-out-of-*n*: G system with repair: The D-policy. *Computers & Operations Research*, *28*(10), 973–981.
13. Grida, M., Zaid, A., Kholief, G. (2017, January). Repairable 3-out-of-4: Cold standby system availability. In *Reliability and Maintainability Symposium (RAMS), 2017 Annual* (pp. 1–6). IEEE.
14. Haggag, M. Y. (2015). Profit analysis and availability of a repairable redundant 3-out-of-4 system involving preventive maintenance. *International Journal of Scientific & Engineering Research*, *6*(8), 1161–1173.

15. Haggag, M. Y. (2009). Cost analysis of k-out of-n repairable system with dependent failure and stand by support using Kolmogorov's forward equations method. *Journal of Mathematics and Statistics*, 5(4), 401.
16. Mishra, A., Jain, M. (2013). Availability of k-out-of-n: F secondary subsystem with general repair time distribution. *IJE Transactions A: Basics*, 26(7), 743–752.
17. Arulmozhi, G. (2002). Reliability of an M-out-of-N warm standby system with R repair facilities. *Opsearch*, 39(2), 77–87.
18. Barron, Y., Frostig, E., Levikson, B. (2006). Analysis of r-out-of-n systems with several repairmen, exponential life times and phase type repair times: An algorithmic approach. *European Journal of Operational Research*, 169(1), 202–225.
19. Krishnan, R., Somasundaram, S. (2012). Reliability and profit analysis of repairable k-out-of-n system with sensor. *European Journal of Scientific Research*, 67(2), 215–222.
20. Choudhury, G. (1996). On a Poisson queue with general setup. *Indian Journal of Pure and Applied Mathematics*, 27(12), 1199–1211.

APPENDIX 6.1 (FORMULATION OF EQUATIONS)

The probability that the system is in the state S_0 in the time interval $(t, t + \Delta t)$ is given by

$$P_0(t + \Delta t) = (1 - \lambda_1 \Delta t)(1 - \lambda_2 \Delta t)(1 - \lambda_3 \Delta t)(1 - \lambda_A \Delta t)(1 - \lambda_B \Delta t)P_0(t) + \mu_1 P_1(t)\Delta t + \mu_2 P_2(t)\Delta t$$

$$+ \mu_3 P_3(t)\Delta t \int_0^\infty \mu_A P_7(x,t)\Delta t\, dx + \int_0^\infty \mu_B P_8(x,t)\Delta t\, dx + \sum_{i=4}^{6} \int_0^\infty \mu P_i(x,t)\Delta t\, dx$$

$$\frac{P_0(t + \Delta t) - P_0(t)}{\Delta t} + (\lambda_1 + \lambda_2 + \lambda_3 + \lambda_A + \lambda_B)P_0(t) = \mu_1 P_1(t) + \mu_2 P_2(t) + \mu_3 P_3(t)$$

$$+ \int_0^\infty \mu_A P_7(x,t)\, dx + \int_0^\infty \mu_B P_8(x,t)\, dx + \sum_{i=4}^{6} \int_0^\infty \mu P_i(x,t)\, dx$$

Now taking $\lim_{\Delta t \to 0}$, we get

$$\Rightarrow \lim_{\Delta t \to 0} \frac{P_0(t + \Delta t) - P_0(t)}{\Delta t} + \left(\lambda_1 + \lambda_2 + \lambda_3 + \lambda_A + \lambda_B\right)P_0(t) = \mu_1 P_1(t) + \mu_2 P_2(t) + \mu_3 P_3(t)$$

$$+ \int_0^\infty \mu_A P_7(x,t)\, dx + \int_0^\infty \mu_B P_8(x,t)\, dx + \sum_{i=4}^{6} \int_0^\infty \mu P_i(x,t)\, dx$$

$$\Rightarrow \left[\frac{\partial}{\partial t} + \lambda_1 + \lambda_2 + \lambda_3 + \lambda_A + \lambda_B\right]P_0(t) = \mu_1 P_1(t) + \mu_2 P_2(t)$$

(A6.1)

$$+ \mu_3 P_3(t) + \int_0^\infty \mu_A P_7(x,t)\, dx + \int_0^\infty \mu_B P_8(x,t)\, dx + \sum_{i=4}^{6} \int_0^\infty \mu P_i(x,t)\, dx$$

For state S_1

$$P_1(t+\Delta t) = (1-\lambda_2\Delta t)(1-\lambda_3\Delta t)(1-\mu_1\Delta t)P_1(t) + \lambda_1 \Delta t P_0(t)$$

$$\frac{P_1(t+\Delta t) - P_1(t)}{\Delta t} + (\lambda_2 + \lambda_3 + \mu_1) P_1(t) = \lambda_1 P_0(t)$$

Now taking $\lim_{\Delta t \to 0}$, we get

$$\lim_{\Delta t \to 0} \frac{P_1(t+\Delta t) - P_1(t)}{\Delta t} + (\lambda_2 + \lambda_3 + \mu_1) P_1(t) = \lambda_1 P_0(t)$$

$$\Rightarrow \left[\frac{\partial}{\partial t} + \lambda_2 + \lambda_3 + \mu_1 \right] P_1(t) = \lambda_1 P_0(t)$$

(A6.2)

For state S_2

$$P_2(t+\Delta t) = (1-\lambda_1\Delta t)(1-\lambda_3\Delta t)(1-\mu_2\Delta t)P_2(t) + \lambda_2 \Delta t P_0(t)$$

$$\frac{P_2(t+\Delta t) - P_2(t)}{\Delta t} + (\lambda_1 + \lambda_3 + \mu_2) P_2(t) = \lambda_2 P_0(t)$$

Now taking $\lim_{\Delta t \to 0}$, we get

$$\lim_{\Delta t \to 0} \frac{P_2(t+\Delta t) - P_2(t)}{\Delta t} + (\lambda_1 + \lambda_3 + \mu_2) P_2(t) = \lambda_2 P_0(t)$$

$$\Rightarrow \left[\frac{\partial}{\partial t} + \lambda_1 + \lambda_3 + \mu_2 \right] P_2(t) = \lambda_2 P_0(t)$$

(A6.3)

For state S_3

$$P_3(t+\Delta t) = (1-\lambda_1\Delta t)(1-\lambda_2\Delta t)(1-\mu_3\Delta t)P_3(t) + \lambda_3 \Delta t P_0(t)$$

$$\frac{P_3(t+\Delta t) - P_3(t)}{\Delta t} + (\lambda_1 + \lambda_2 + \mu_3) P_3(t) = \lambda_3 P_0(t)$$

Now taking $\lim_{\Delta t \to 0}$, we get

$$\lim_{\Delta t \to 0} \frac{P_3(t+\Delta t) - P_3(t)}{\Delta t} + (\lambda_1 + \lambda_2 + \mu_3) P_3(t) = \lambda_3 P_0(t)$$

$$\Rightarrow \left[\frac{\partial}{\partial t} + \lambda_1 + \lambda_2 + \mu_3 \right] P_3(t) = \lambda_3 P_0(t)$$

(A6.4)

For state S_4

$$P_4(x+\Delta x, t+\Delta t) = (1-\mu\Delta t)P_4(x,t)$$

$$\lim_{\substack{\Delta x \to 0 \\ \Delta t \to 0}} \frac{P_4(x+\Delta x, t+\Delta t) - P_4(x,t)}{\Delta t} + \mu P_4(x,t) = 0$$

$$\Rightarrow \left[\frac{\partial}{\partial x} + \frac{\partial}{\partial t} + \mu \right] P_4(x,t) = 0$$

(A6.5)

For state S_5

$$P_5(x+\Delta x,t+\Delta t)=(1-\mu\Delta t)P_5(x,t)$$

$$\lim_{\substack{\Delta x\to 0\\\Delta t\to 0}}\frac{P_5(x+\Delta x,t+\Delta t)-P_5(x,t)}{\Delta t}+\mu P_5(x,t)=0$$

(A6.6)

$$\Rightarrow\left[\frac{\partial}{\partial x}+\frac{\partial}{\partial t}+\mu\right]P_5(x,t)=0$$

For state S_6

$$P_6(x+\Delta x,t+\Delta t)=(1-\mu\Delta t)P_6(x,t)$$

$$\lim_{\substack{\Delta x\to 0\\\Delta t\to 0}}\frac{P_6(x+\Delta x,t+\Delta t)-P_6(x,t)}{\Delta t}+\mu P_6(x,t)=0$$

(A6.7)

$$\Rightarrow\left[\frac{\partial}{\partial x}+\frac{\partial}{\partial t}+\mu\right]P_6(x,t)=0$$

For state S_7

$$P_7(x+\Delta x,t+\Delta t)=(1-\mu_A\Delta t)P_7(x,t)$$

$$\lim_{\substack{\Delta x\to 0\\\Delta t\to 0}}\frac{P_7(x+\Delta x,t+\Delta t)-P_7(x,t)}{\Delta t}+\mu_A P_7(x,t)=0$$

(A6.8)

$$\Rightarrow\left[\frac{\partial}{\partial x}+\frac{\partial}{\partial t}+\mu_A\right]P_7(x,t)=0$$

For state S_8

$$P_8(x+\Delta x,t+\Delta t)=(1-\mu_B\Delta t)P_8(x,t)$$

$$\lim_{\substack{\Delta x\to 0\\\Delta t\to 0}}\frac{P_8(x+\Delta x,t+\Delta t)-P_8(x,t)}{\Delta t}+\mu_B P_8(x,t)=0$$

(A6.9)

$$\Rightarrow\left[\frac{\partial}{\partial x}+\frac{\partial}{\partial t}+\mu_B\right]P_8(x,t)=0$$

APPENDIX 6.2 (BOUNDARY CONDITIONS)

The following boundary conditions are also obtained from the following:

$$P_4(0,t)=\lambda_3 P_1(t)+\lambda_1 P_3(t)$$

(A6.10)

$$P_5(0,t)=\lambda_1 P_2(t)+\lambda_2 P_1(t)$$

(A6.11)

$$P_6(0,t) = \lambda_3 P_2(t) + \lambda_2 P_3(t) \tag{A6.12}$$

$$P_7(0,t) = \lambda_A P_0(t) \tag{A6.13}$$

$$P_8(0,t) = \lambda_B P_0(t) \tag{A6.14}$$

Initial condition

$$P_i(t) = \begin{cases} 1, & t = 0 \text{ and } i = 0 \\ 0, & \text{otherwise} \end{cases} \tag{A6.15}$$

7 Minimizing the Discounted Cash Flow of Weighted Delay Penalty Costs in Resource-Constrained Project Scheduling Problems Using Three Metaheuristic Algorithms

Mohammad Khalilzadeh
CENTRUM Católica Graduate Business School, Lima, Peru
Pontificia Universidad Católica del Perú, Lima, Peru

Sayyid Ali Banihashemi
Department of Industrial Engineering,
Payame Noor University, Tehran, Iran

CONTENTS

DOI: 10.1201/9781003156291-7

7.1 INTRODUCTION

Project scheduling is of extremely important in the field of project management and optimization. Various companies seek to utilize project management to reduce project duration and cost, and increase quality and customer satisfaction. Project scheduling problem (PSP) deals with the determination of a schedule for performing a set of related activities that constitute a project. The schedule should be developed in such a way as to optimize the objective function along with meeting the precedence relationships and resource constraints. The successful accomplishment of a project is when all project activities are completed on time.

A classification of project scheduling projects that does not take resource constraints into account has been known as non-resource-constrained problem (NRCPSP), and the other that considers resource limitations and availability has been recognized as resource-constrained project scheduling problem (RCPSP). Relating to the allocated resources, a project activity may be carried out in two or more different modes. In addition, each activity may be interrupted during execution. Activity precedence relationships and resource availability can also be classified into different categories. Finally, there are various objective functions in project scheduling, the most important of which include minimizing project duration, maximizing net present value (NPV), minimizing the total cost, minimizing the fluctuations of resource consumption (leveling resources), and maximizing quality level. Several project scheduling models can be defined considering various assumptions related to the different actual conditions of projects.

Project scheduling is of significant importance from practical and theoretical points of view. In practice, a proper project schedule leads to increasing benefits and stakeholders' satisfaction. Regarding the theoretical viewpoint, project scheduling is a fascinating research topic for scholars, especially operations research specialists. In recent years, many new models and algorithms have been developed and implemented in the field of project scheduling; however, there are still many shortcomings in this regard. Despite numerous studies in various PSPs, there are still many potential areas for further research. Existing shortcomings are in two aspects: developing models closer to the real-world conditions and improving solution methods. Regarding the first aspect, despite considering numerous assumptions for the PSPs, it is still possible to present new models. It is obvious that considering the assumptions should be in accordance with real-world conditions.

Regarding the second point of view, new tools and techniques are always provided for solving problems as a result of the progress of science in areas related to project scheduling. It should be noted that because of the NP-Hardness of RCPSP (Blazewicz et al., 1983), it is impossible to solve large-size problems with exact methods in a reasonable time. Therefore, several heuristic and metaheuristic approaches have been presented to solve these types of optimization problems. These methods can obtain close to optimal solutions within a reasonable time. In this study, a new classification of resource-constrained PSPs is introduced and solved by using three different metaheuristic algorithms. The objective function of the proposed model is to minimize the discounted cash flow of the weighted delay penalty costs in the usage of renewable resources. In this type of RCPSP that is applicable to numerous real-world projects, it is supposed that renewable (nonconsumable) resources are limited and expensive and can be used in other projects. In other words, they must be rented and cannot be accessible all the time of project implementation. Hence, there is a predetermined time for accessing these renewable resources as well as a deadline for returning them so that no resources are available before that predetermined time; however, after the deadline, these resources can still be accessible by paying penalty cost depending on their type. The metaheuristic algorithms that are exploited to deal with this problem include imperialist competitive algorithm (ICA), biogeography-based optimization (BBO), and shuffled frog-leaping algorithm (SFLA). This proposed problem is solved by using these algorithms several times, and the obtained results are compared with each other.

7.2 LITERATURE REVIEW

In 1951, the primary models and project scheduling methods were introduced for large-sized projects. The critical path method (CPM) was first proposed by Kelley in 1961 considering deterministic activity duration and finish-to-start (FS) activity precedence relationship. However, the required resources for performing project activities are not taken into account in the CPM method. It should be noted that the project delay is mainly due to the limited number of available renewable resources as well as the delay in their timely delivery.

Bianco et al. (1998) considered a type of PSP in which resources can only be allocated to one activity at a given time period. Pinedo and Singer (1999) investigated unlike branch and bound (B&B) methods to solve the problem with the objective function of minimizing total weighted delays. Vanhoucke et al. (2001) developed a B&B method together with an exact recursive search method to solve the RCPSP problem with the objective of minimizing the total earliness-tardiness. Demeulemeester and Herroelen (2002) introduced a B&B method for solving the PSP with the objective of minimizing weighted earliness-tardiness (RCPSPWET) problem. Afshar-Nadjafi and Shahram Shadrokh (2009) proposed a B&B method for solving a type of PSP with the total weighted earliness-tardiness considering general activity precedence relationships. However, they did not take the resource limitation into account. Lova et al. (2006) used a serial scheduling scheme for the resource-constrained RCPSPWET. They showed that the serial scheduling scheme usually outperforms the parallel scheduling scheme. In addition, Lova et al. (2009) employed a hybrid genetic algorithm (GA) for solving the RCPSPWET problem. Chen et al. (2010) used a delay local research method to escape from getting trapped in the local optimal solutions. This method is similar to the mutation operator in the GA algorithm and can postpone the scheduling of each activity despite the priority of that activity to allow other activities to be scheduled earlier. Khalilzadeh et al. (2012) exploited the particle swarm optimization (PSO) algorithm to solve the RCPSP problem with the goal of minimizing total weighted earliness-tardiness costs and also solved the same problem of minimizing the total costs of renewable and nonrenewable resources in a fuzzy environment using the modified PSO algorithm. Golestaneh et al. (2013) used the GA and Tabu Search (TA) algorithms to tackle the RCPSP problem with the objective of minimizing delay penalty costs.

Several studies have examined RCPSP problems so far. Some examples are brought as follows. Afshar-Nadjafi (2018) studies the preemptive multimode RCPSP problem with the goal of minimizing the project makespan considering mode changeability after activity preemption. Creemers (2019) examined the preemptive stochastic RCPSP problem and presented an exact solution procedure. Vanhoucke and Coelho (2019) proposed a solution algorithm to solve RCPSP problem considering split able activities and setup time. Maghsoudlou et al. (2020) suggested a framework for preemptive multiskilled RCPSP problem with time-of-use energy tariffs. Khalilzadeh (2021) developed an optimization model for RCPSP problem with the objective function of resource leveling taking preemptive and multimode activities as well as renewable and nonrenewable resources into account. But, none of them considered earliness-tardiness issue as a prevailing problem in any project.

Despite the necessity and importance of the RCPSPWET problem in real-world projects, very limited studies have examined this type of RCPSP problem so far. In the current study, the RCPSP problem is investigated taking the time value of money into account.

7.3 MATERIALS AND METHODS

7.3.1 MATHEMATICAL PROGRAMMING MODEL

Each project comprises n ($i=1,2,3,...,n$) activities. It is supposed that R number of renewable (nonconsumable) resources ($k=1,2,3,...,R$) are required to implement the activities. The length and duration of implementing activity j is denoted by d_j. Dummy (artificial) beginning and end activities are considered to show the launch and finish time of the project. r_{kj} shows the necessary amount of renewable (nonconsumable) resource k for implementing activity j. R_k indicates the availability of resource-type k throughout the project horizon. It is assumed that the precedence relationship between activities is FS. In addition, the project activities cannot be interrupted. The decision variable is shown as s_j which displays the start (beginning) time of project activity j. The project network is shown by Activity-On-Arc (AON) graph. ρ_k, δ_k, T_k, and P_k show the ready time of resource k, time to return resource k, the amount of delay time in returning resource k, and the penalty cost of delay in returning resource k, respectively.

Since the time value of money is always considered in real-world projects, the NPV is taken into account using discounted cash flows. A continuous discount rate of α is denoted by α^T. It should be noted that the only objective function of the proposed mathematical programming model is to minimize the discounted cash flows corresponding with the total penalty costs of delay in returning the renewable resources. This RCPSP problem can be modeled and formulated as follows. The notations are represented in Table 7.1.

Objective function:

$$min\ Z = \sum_{k=1}^{R} \alpha^T\ P_k T_k \tag{7.1}$$

Constraints:

$$\rho_k \leq s_i \qquad i \in N_k,\ k=1,2,3,...,R \tag{7.2}$$

$$T_k \geq CP_k - \delta_k \quad k=1,2,3,...,R \tag{7.3}$$

TABLE 7.1
Notations

Indices:

n	The number of project activities
R	The number of renewable resources
P_j	The set of the predecessors of activity j
t	Time period of t

Parameters:

d_j	Duration of activity j
R_k	Total available amount of resource k
r_{kj}	Requisite amount of renewable (nonconsumable) resource k for implementing activity j
ρ_k	Ready time of resource k
δ_k	Deadline for returning resource k
T_k	Amount of delay time in returning resource k
P_k	Penalty cost of delay in returning resource k
∞^T	Continuous discount rate of α
s_i	Start (beginning) time of activity i
s_j	Start (beginning) time of activity j
N_k	The set of activities that need resource k
P_j	The set of predecessor of activity j
CP_k	Time of returning resource k

$$T_k \geq 0 \qquad k = 1,2,3,\ldots,R \qquad (7.4)$$

$$s_j - s_i \geq d_i \qquad i \in P_j,\ j = 1,2,3,\ldots,n \qquad (7.5)$$

$$\sum_{j \in p(t)} r_{jk} \leq R_k \qquad k = 1,2,3,\ldots,R \qquad (7.6)$$

$$CP_k, s_i, s_j \geq 0 \qquad \begin{aligned} i &= 0,1,2,3,\ldots,n+1 \\ j &= 1,2,3,\ldots,n \end{aligned} \qquad (7.7)$$

Equation (7.1) represents the goal and objective function of the proposed optimization model, which seeks to minimize the total weighted penalty costs of returning the renewable resources. Equation (7.2) dictates that the resource-type k must be ready before starting activity i that requires renewable resource-type k. Equation (7.4) describes the tardiness in returning resource k which is the difference between the deadline and time of returning resource k. Equation (7.4) mandates that T_k must be greater than or equal to zero. Equation (7.5) denotes the precedence relationships. Equation (7.6) dictates that the total amount of usage of resource k per unit of time cannot exceed the total available amount of resource k. Finally, Equation (7.7) defines the decision variables of the proposed model.

7.3.2 Solution Methodology

Generally, the methods for solving the RCPSP problems are categorized into three different groups, including exact methods, heuristic, and metaheuristic algorithms. The B&B procedures together with the metaheuristic algorithms are the most frequently used methods for solving the RCPSP problems. As mentioned above, Blazewicz et al. (1983) demonstrated that the RCPSP problems are NP-Hard. As a result, the exact methods such as B&B are not able to find the optimal solutions for large-size RCPSP problems in a reasonable computational time. Therefore, meta-heuristic algorithms can be considered as reasonable solution methods for dealing with the RCPSP problems. In this study, three metaheuristic algorithms named ICA, BBO algorithm, and SFLA are employed to deal with the RCPSPWET problem considering the time value of money.

7.3.2.1 Imperialist Competitive Algorithm (ICA)

What is remarkable is that most of the metaheuristic optimization algorithms are simulations of natural processes. Perhaps one reason can be the simplicity of formulating and understanding the evolution of these processes. In this section, an algorithm inspired by sociopolitical processes has been considered for optimization. The ICA algorithm has been known as a metaheuristic algorithm that has higher capability and speed than many other metaheuristic algorithms presented so far. Figure 7.1 presents an overview of the ICA algorithm (Atashpaz-Gargari and Lucas, 2007).

The ICA algorithm, like other evolutionary optimization methods, initiates with a number of initial populations. In this algorithm, each member of the population

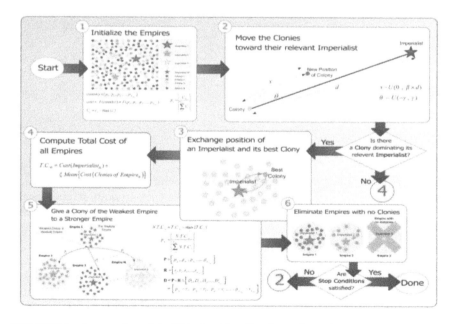

FIGURE 7.1 An overview of the ICA algorithm (Atashpaz-Gargari and Lucas, 2007).

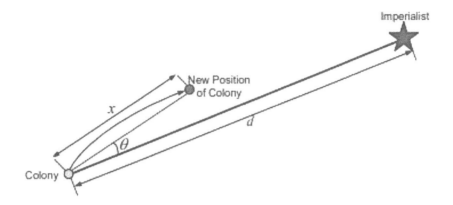

FIGURE 7.2   The movement of the colonies toward the imperialist (Atashpaz-Gargari and Lucas, 2007).

is called a country. Countries are in fact possible solutions to the problem and are equivalent to chromosomes in the GA and PSO algorithm. Countries are divided into two categories of colonies and colonizers. Each colonizer, depending on its power, dominates and controls a number of colonial countries. Colonial absorption and competition policies form the core of this algorithm. This policy is carried out by moving the colonies of an imperialist. Figure 7.2 shows this movement (Atashpaz-Gargari and Lucas, 2007).

If a colony attains a better position than the colonizer on the move, they will be swapped. Meanwhile, the total power of an empire is defined as the total power of the colonial country plus a percentage of the average power of its colonies. Colonial competition is another important part of this algorithm. During colonial rivalry, weak empires gradually lose their power and weaken over time. Colonial rivalry leads to only one empire in the world. This is when the ICA algorithm reaches the optimal point and stops. Figure 7.3 displays the overall scheme of imperialist competition (Atashpaz-Gargari and Lucas, 2007).

Figure 7.4 presents the flowchart of this algorithm.

To initiate the algorithm, $N_{country}$ number of initial countries is created. From the best N_{imp} number of this population, the countries with the lowest cost objective function values are selected as imperialists. The remaining N_{col} number of the countries forms the colonies each of which belongs to an imperialist. To divide the initial colonies among the imperialists, each imperialist is given a number of colonies, the number of which is proportional to its power. The normalized values of imperialists are calculated as follows:

$$C_n = \max_{i=1,\ldots,N_{imp}} \{c_i\} - c_n \qquad (7.8)$$

where c_n is the cost objective function value of the imperialist n, $\max_{i=1,\ldots,N_{imp}} \{c_i\}$ is the maximum cost objective function value among the imperialists, and C_n is the

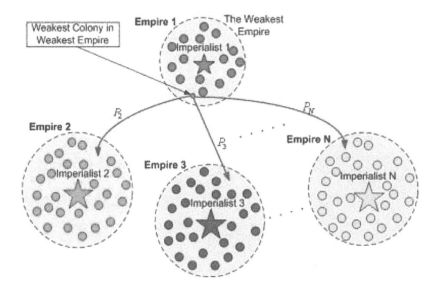

FIGURE 7.3   Overall scheme of imperialist competition (Atashpaz-Gargari and Lucas, 2007).

normalized value of the imperialist n. The imperialist that has a higher cost objective function value has a lower normalized value. Given the normalized value, the relative normalized power of each imperialist is calculated as follows:

$$p_n = \left| \frac{C_n}{\sum_{i=1}^{N_{imp}} C_i} \right| \qquad (7.9)$$

On the other hand, the normalized power of an imperialist is the proportion of the colonies ruled by that imperialist. As a result, the initial number of colonies of an imperialist would be equal to:

$$NC_n = round\{p_n N_{col}\} \qquad (7.10)$$

where NC_n is the initial number of colonies of imperialist n, N_{col} is the total number of colonies. *round* is also a function that gives the closest integer to a decimal number. For more information, please read Atashpaz-Gargari and Lucas (2007). Figure 7.5 depicts how the initial empires form.

7.3.2.2   Biogeography-Based Optimization (BBO) Algorithm

BBO algorithm, developed by Simon, is a new global optimization algorithm based on biogeography theory, which studies the geographical distribution of biological organisms. Mathematical biogeographical models describe how species migrate from one island to another, how new species emerge, and how species become extinct. Geographical areas that are suitable habitats for biodiversity have a habitat suitability index (HSI). Sudden changes can change the HSI value of a habitat. The higher the

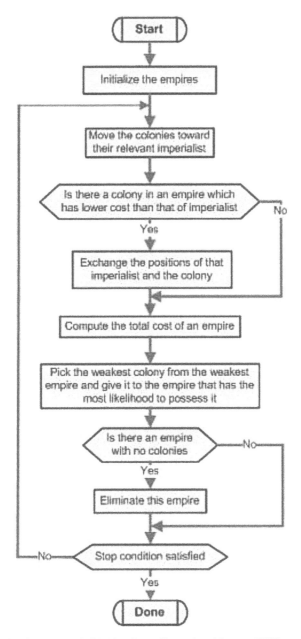

FIGURE 7.4   The flowchart of ICA (Atashpaz-Gargari and Lucas, 2007).

HSI, the better the location – however, for the minimization problem, the less the better. If there is a species in a location that wants to leave the site, it is more likely that this species is from the location with the higher population because it has many competitors and the location with the less population has more potential for migration, which is represented by Figure 7.6 and Equation (7.11).

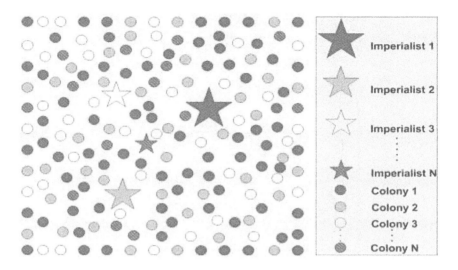

FIGURE 7.5   How initial empires form (Atashpaz-Gargari and Lucas, 2007).

In Figure 7.6, *I* and *E* represent the highest possible immigration rate and the highest possible migration rate, respectively. λ and μ denote immigration rate and migration rate of habitat, respectively. S_0 and S_{max} indicate the equilibrium point and the maximum possible number of species in the habitat, respectively.

$$\mu + \lambda = \lambda max = \mu max \qquad (7.11)$$

7.3.2.3   Shuffled Frog-Leaping Algorithm (SFLA)

The SFLA algorithm introduced by Eusuff et al. (2006) is a memetic-based meta-heuristic algorithm inspired by frog groups searching for food. The researchers found that individual group members could use the findings and previous experiences of other group members while searching for food. When food resources are unpredictably distributed in different parts, the advantages of this method outweigh the disadvantages of competition for food.

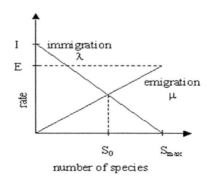

FIGURE 7.6   Number of species vs. rate.

This algorithm employs the evolving mimetic method for local search among frogs' subgroups. It uses a combination strategy and allows the exchange of messages in local search. Messages are exchanged not only in local search but also in global search. Thus, in this algorithm, local and global searches are well combined. This algorithm initiates with a random choice of frog groups. Frog groups are classified into different subgroups, each of which can independently and differently do local search. A subgroup's frogs can influence on the generations of frogs in that given subgroup. This is how a subgroup's frogs evolve. Each frog has its own culture and can use the cultures or ideas of other frogs during evolution.

Memetic evolution improves the memetic quality of individual frogs and increases the ability to achieve the goal. Regarding the objective function, good frogs' weights can be increased and bad frogs' weights can be reduced. The balance between the exchange of global message and local search allows the algorithm to easily escape from the local minimum and achieve global optimization using the Shuffling strategy. Fast convergence is one of the privileges and advantages of this algorithm. The first step of the algorithm is to generate the initial solutions. Producing appropriate initial solutions plays an important role in the convergence speed and accuracy of the algorithm. The initial population is randomly generated. Each frog updates its position based on the best-achieved solution so far. Because of the effect that the best local solution has on the frog, all the frogs will quickly gather around it.

SFLA begins with the initial population consisting of P number of frogs, which is randomly generated from the feasible space Ω. This population is divided into m groups, shown in Figure 7.7.

The process of dividing frogs is such that the first frog belongs to the first set, the second frog belongs to the second set, and the frog m belongs to the m set, and the frog $m+1$ belongs to the first set. Each set contains n frogs so that the following relation can be written.

$$P = m \times n \qquad (7.12)$$

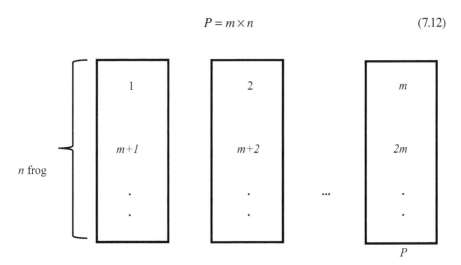

FIGURE 7.7   Frog classification.

In the D-dimensional problems, the position of the ith frog is displayed as $(x_{i1}, x_{i2}, x_{i3}, \ldots, x_{iD})$. Each frog is then ranked according to its position, and the frogs are sorted in descending order based on their fitness values. The frogs that have the best and worst fitness values are displayed with x_b and x_w, respectively. Also, the frog that has the best fitness value in the whole population is displayed with x_g. Subsequently, using an interactive process, the merits of the worst frogs in each cycle of the algorithm (not all frogs) are corrected. During the evolutionary process, each frog corrects its position in a way that brings it closer to the best frog. This correction is done as follows:

$$Step\ size\ s_i = rand(\ ) \times (x_b - x_w);$$
$$New\ position\ x_w = current\ position\ x_w + s_i; \tag{7.13}$$
$$-s_{max} \leq s_i \leq s_{max}$$

where $rand(\ )$ is a random number between 0 and 1. s_{max} is the maximum virtual amount that a frog can jump after being affected by other frogs. If a better solution is produced during this process, that better solution will replace the worst solution (frog). Otherwise, the calculations are redone using the above equations according to the best frog in the whole population. In other words, x_b is replaced by x_g. If no improvement is made in this case, a new solution is randomly created to replace the worst frog, and then these calculations are repeated a certain number of times.

7.4　RESULTS AND DISCUSSION

Initially, the proposed model and metaheuristic algorithms are validated on a small-sized problem. It is shown that each of these algorithms can provide close to the optimal solutions for the problem in hand. Then, the outputs of these three algorithms are examined in terms of solution quality and CPU computational time. For this purpose, the problems are divided into three categories of 30 (small size), 60 (medium size), and 90 (large size), and the algorithms are separately analyzed based on each category.

7.4.1　ALGORITHM VALIDATION

In this section, a project, including 7 activities, was considered to validate the proposed model and metaheuristic algorithms. Dummy activities 0 and 8 indicate the start and finish times of the project, shown in Figure 7.8. The activities that require resource-types A and B are also shown in Figure 7.8.

In addition, the project data are given in Table 7.2.

Additional data are provided in Table 7.3.

The aforementioned problem was solved using the GAMS software (exact method) as well the three metaheuristic algorithms of ICA, BBO, and SFLA. The results are presented in Table 7.4.

As can be seen, all three metaheuristic algorithms can find the optimal solution for this problem. The only difference between these algorithms is their run time. The shortest computational time belongs to ICA and the longest computational time belongs to SFLA.

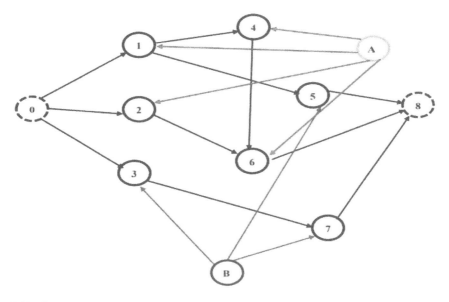

FIGURE 7.8   Project network.

Then, five project instances with 30 activities were considered to compare the performance of these algorithms based on small-size problems. The parameters of these numerical examples were generated using MATLAB software based on continuous uniform distribution. The data of these problems are displayed in Table 7.5.

The available number of each resource is between 7 and 11 units, and the consumption rate of each resource is randomly chosen between 1 and 3 units. After

---

TABLE 7.2
The Project Data

Resource Type	Availability	Ready Time	Delivery Time	Delay Penalty Cost	Activities Requiring Resource
A	4	2	11	14	1,2,4,6
B	5	3	10	11	3,5,7

TABLE 7.3
Additional Data

Activity	1	2	3	4	5	6	7
Duration	2	3	1	4	2	1	3
The Required Amount of Resource-Type A	2	4	–	2	–	6	–
The Required Amount of Resource-Type B	–	–	3	–	5	–	3

TABLE 7.4

Comparison of the Computational Results

GAMS		ICA		BBO		SFLA	
Solution obtained (objective function value)	Solution obtained (objective function value)	Computational CPU time (second)	Solution obtained (objective function value)	Computational CPU time (second)	Solution obtained (objective function value)	Computational CPU time (second)
14	14	0.19	14	0.27	14	0.29

randomly generating the numerical examples, each project instance was solved using the metaheuristic algorithms with the MATLAB software. Each problem was run 10 times. The findings are shown in Table 7.6.

In Table 7.6, *min* represents the best solution obtained by this algorithm. *Average* depicts the average objective function value for each problem in 10 times of the algorithm run. *Time* displays the average computational time of the algorithm (in seconds) for each problem in 10 times of the algorithm run. As can be seen in Table 7.5, the average computational times of ICA, BBO, and SFLA are 48.52, 48.5, and 53.32 seconds, which demonstrates that the ICA algorithm has the shortest average computational time. Figure 7.9 exhibits the average computational time of each algorithm.

However, the solution quality is important. Figure 7.10 shows the best solution obtained by each proposed algorithm.

As it can be seen, the ICA algorithm performs worse than the other two algorithms. The best solutions were obtained by the SFLA algorithm for problem 1 (P30_1) and problem 3 (P30_3). However, the BBO algorithm found the best solutions for the other problems.

TABLE 7.5

The Problem Data

Problem Number	Activity Duration	Number of Resources	Delay Penalty cost	Ready Time	Delivery Time
P30_1	$U(1,5)$	2	$U(10,30)$	$U(0,2)$	$U\left(\dfrac{\sum_{i=1}^{n} t_i}{5}, \dfrac{\sum_{i=1}^{n} t_i}{3}\right)$
P30_2	$U(1,5)$	3	$U(10,30)$	$U(0,2)$	$U\left(\dfrac{\sum_{i=1}^{n} t_i}{5}, \dfrac{\sum_{i=1}^{n} t_i}{3}\right)$
P30_3	$U(1,5)$	4	$U(10,30)$	$U(0,2)$	$U\left(\dfrac{\sum_{i=1}^{n} t_i}{5}, \dfrac{\sum_{i=1}^{n} t_i}{3}\right)$
P30_4	$U(1,5)$	5	$U(10,30)$	$U(0,2)$	$U\left(\dfrac{\sum_{i=1}^{n} t_i}{5}, \dfrac{\sum_{i=1}^{n} t_i}{3}\right)$
P30_5	$U(1,5)$	6	$U(10,30)$	$U(0,2)$	$U\left(\dfrac{\sum_{i=1}^{n} t_i}{5}, \dfrac{\sum_{i=1}^{n} t_i}{3}\right)$

TABLE 7.6

The Findings of Solving the Small-Size Problems with the Three Metaheuristic Algorithms

	ICA			BBO			SFLA		
	Min	Average	Time (s)	Min	Average	Time (s)	Min	Average	Time (s)
P30_1	169	181	29.3	167	177	29.4	165	174	34.8
P30_2	304	327	35.1	292	317	33.8	299	315	39.7
P30_3	348	361	44.8	348	358	48.9	340	355	47.9
P30_4	537	595	60.1	522	594	61.1	537	586	69.4
P30_5	619	640	73.3	597	620	69.3	607	620	74.8
Average			48.52			48.5			53.32

Five numerical examples with 60 activities were considered to compare the performance of the algorithms for medium-size projects. The parameters of these numerical examples were generated using MATLAB software based on continuous uniform distribution. The data of these problems are displayed in Table 7.7.

The available number of each resource is between 7 and 11 units, and the consumption rate of each resource is randomly chosen between 1 and 3 units. After randomly generating the numerical examples, each project instance was solved using the metaheuristic algorithms with the MATLAB software. Each problem was run 10 times. The findings are presented in Table 7.8.

As it can be seen, the ICA algorithm has the shortest computational time. However, the BBO algorithm found the best solutions for all five medium-size problems. Figure 7.11 displays the average computational time of each algorithm.

In addition, Figure 7.12 shows the best solution obtained by each proposed algorithm.

Five numerical examples with 90 activities were considered to evaluate and compare the efficiency and performance of the algorithms for large-size projects. The parameters

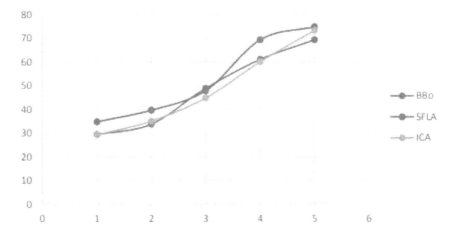

FIGURE 7.9 The average computational time of each algorithm for small-size problems.

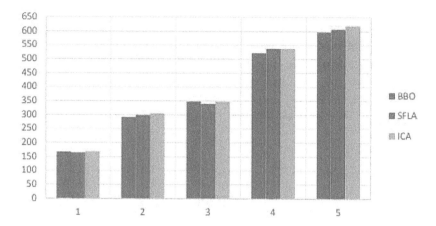

FIGURE 7.10 The graph of the best solution found by each algorithm for small-size problems.

of these numerical examples were generated using MATLAB software based on continuous uniform distribution. The data of these problems are displayed in Table 7.9.

The available number of each resource is between 7 and 11 units, and the consumption rate of each resource is randomly chosen between 1 and 3 units. After randomly generating the numerical examples, each project instance was solved using the metaheuristic algorithms with the MATLAB software. Each problem was run 10 times. The results are presented in Table 7.10.

As it can be seen, the ICA algorithm has the shortest computational time. However, the BBO algorithm found the best solutions for all five large-size problems. Figure 7.13 displays the average computational time of each algorithm.

In addition, Figure 7.14 shows the best solution obtained by each proposed algorithm.

TABLE 7.7
The Problem Data

Problem Number	Activity Duration	Number of Resources	Delay Penalty Cost	Ready Time	Delivery Time
P60_1	$U(1,5)$	2	$U(10,30)$	$U(0,2)$	$U\left(\frac{\sum_{i=1}^{n} t_i}{5}, \frac{\sum_{i=1}^{n} t_i}{3}\right)$
P60_2	$U(1,5)$	3	$U(10,30)$	$U(0,2)$	$U\left(\frac{\sum_{i=1}^{n} t_i}{5}, \frac{\sum_{i=1}^{n} t_i}{3}\right)$
P60_3	$U(1,5)$	4	$U(10,30)$	$U(0,2)$	$U\left(\frac{\sum_{i=1}^{n} t_i}{5}, \frac{\sum_{i=1}^{n} t_i}{3}\right)$
P60_4	$U(1,5)$	5	$U(10,30)$	$U(0,2)$	$U\left(\frac{\sum_{i=1}^{n} t_i}{5}, \frac{\sum_{i=1}^{n} t_i}{3}\right)$
P60_5	$U(1,5)$	6	$U(10,30)$	$U(0,2)$	$U\left(\frac{\sum_{i=1}^{n} t_i}{5}, \frac{\sum_{i=1}^{n} t_i}{3}\right)$

TABLE 7.8

The Results of Solving the Medium-Size Problems with the Three Metaheuristic Algorithms

	ICA			BBO			SFLA		
	Min	*Average*	*Time (s)*	*Min*	*Average*	*Time (s)*	*Min*	*Average*	*Time (s)*
P60_1	430	469	44.5	395	428	58.9	428	489	53.4
P60_2	591	664	60.3	549	611	72.1	561	631	67.5
P60_3	733	782	79.5	704	754	89.3	733	795	82.4
P60_4	917	935	91.3	864	909	95.7	907	928	103.8
P60_5	1119	1130	109.4	1076	1128	99.3	1103	1139	133.4
Average			77			83.06			88.1

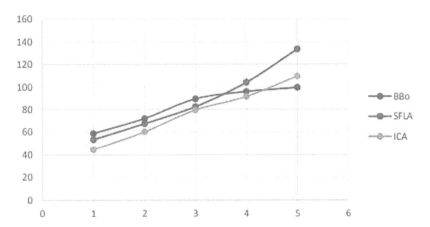

FIGURE 7.11   The average computational time of each algorithm for medium-size problems.

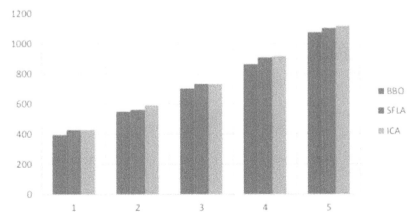

FIGURE 7.12   The graph of the best solution found by each algorithm for medium-size problems.

TABLE 7.9
The Problem Data

Problem Number	Activity Duration	Number of Resources	Delay Penalty Cost	Ready Time	Delivery Time
P90_1	$U(1,5)$	2	$U(10,30)$	$U(0,2)$	$U\left(\dfrac{\sum_{i=1}^{n} t_i}{5}, \dfrac{\sum_{i=1}^{n} t_i}{3}\right)$
P90_2	$U(1,5)$	3	$U(10,30)$	$U(0,2)$	$U\left(\dfrac{\sum_{i=1}^{n} t_i}{5}, \dfrac{\sum_{i=1}^{n} t_i}{3}\right)$
P90_3	$U(1,5)$	4	$U(10,30)$	$U(0,2)$	$U\left(\dfrac{\sum_{i=1}^{n} t_i}{5}, \dfrac{\sum_{i=1}^{n} t_i}{3}\right)$
P90_4	$U(1,5)$	5	$U(10,30)$	$U(0,2)$	$U\left(\dfrac{\sum_{i=1}^{n} t_i}{5}, \dfrac{\sum_{i=1}^{n} t_i}{3}\right)$
P90_5	$U(1,5)$	6	$U(10,30)$	$U(0,2)$	$U\left(\dfrac{\sum_{i=1}^{n} t_i}{5}, \dfrac{\sum_{i=1}^{n} t_i}{3}\right)$

In general, the results show that the ICA algorithm is faster than the other two algorithms in terms of computational time. However, in terms of solution quality, it could not compete with the other two algorithms. On the other hand, the BBO algorithm was able to obtain the solutions with less objective function values than the solutions found by the other two algorithms.

7.5 CONCLUSION

In this study, a new classification of RCPSP problems was investigated considering the minimization of the discounted cash flow of weighted delay penalty costs as the objective function of the problem. Due to the NP-Hardness of the problem in hand, three metaheuristic algorithms named ICA, BBO, and SFLA were applied

TABLE 7.10
The Results of Solving the Large-Size Problems with the Three Metaheuristic Algorithms

	ICA			BBO			SFLA		
	Min	Average	Time (s)	Min	Average	Time (s)	Min	Average	Time (s)
P60_1	1515	1560	133.9	1427	1697	134.9	1504	1668	148.2
P60_2	1680	1948	189.8	1675	1740	186.2	1680	1847	219.4
P60_3	1925	1994	207.3	1906	1970	219.7	1925	1962	228.9
P60_4	2334	2395	231.9	2309	2348	244.9	2334	2375	241.8
P60_5	–	–	–	–	–	–	–	–	–
Average			190.725			196.425			209.575

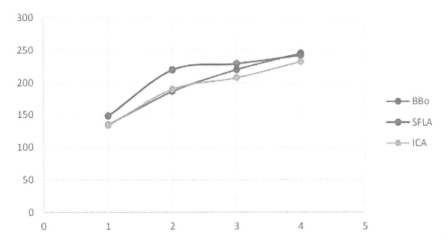

FIGURE 7.13 The average computational time of each algorithm for large-size problems.

to deal with the problem. Various numerical examples, including 30, 60, and 90 activities, were taken into consideration to evaluate the performance of these three metaheuristic algorithms. The parameters of these project instances were randomly generated using MATLAB software based on continuous uniform distribution. The results demonstrated that the ICA algorithm has the shortest computational CPU time compared to the other two metaheuristic algorithms; however, the BBO algorithm performed better than the others in terms of solution quality.

As some suggestions for further studies, the proposed model can be applied to real-world projects. Also, uncertain parameters such as time, cost, and resource availability should be considered. Moreover, other metaheuristic algorithms may be exploited and the results are compared.

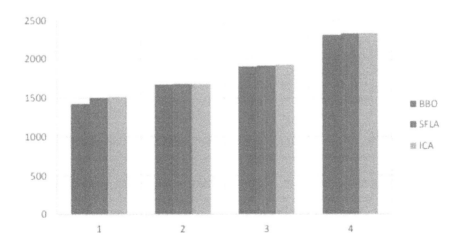

FIGURE 7.14 Graph of the best solution found by each algorithm for large-size problems.

REFERENCES

Afshar-Nadjafi, B. (2018). A solution procedure for preemptive multi-mode project scheduling problem with mode changeability to resumption. *Applied Computing and Informatics, 14(2)*, pp. 192–201.

Atashpaz-Gargari, E., and Lucas C. (2007). Imperialist competitive algorithm: An algorithm for optimization inspired by imperialistic competition. *2007 IEEE Congress on Evolutionary Computation*, pp. 4661–4667. https://doi:10.1109/CEC.2007.4425083.

Afshar-Nadjafi, B., and Shahram Shadrokh, S. (2009). A branch and bound algorithm for the weighted earliness-tardiness project scheduling problem with generalized precedence relations. *Scientia Iranica, 16(1)*, pp. 55–64.

Bianco, L., Dell'Olmo, and Grazia Speranza, M. (1998). Heuristics for multimode scheduling problems with dedicated resources. *European Journal of Operational Research, 107(2)*, pp. 260–271. https://doi.org/10.1016/S0377-2217(97)00347-0.

Blazewicz, J., Lenstra, J.K., and RinnooyKan, A.H.G. (1983). Scheduling subject to resource constraints: classification and complexity. *Discrete Applied Mathematics, 5(1)*, pp. 11–24. https://doi.org/10.1016/0166-218X(83)90012-4.

Chen, C-Y., Lin, J-W., Lee, W-I., and Chen, C-W. (2010). Fuzzy control for an oceanic structure: a case study in time-delay TLP system. *Journal of Vibration and Control, 16(1)*, pp. 147–160. https://doi:10.1177/1077546309339424.

Creemers, S. (2019). The preemptive stochastic resource-constrained project scheduling problem. *European Journal of Operational Research, 277(1)*, pp. 238–247.

Demeulemeester, E., and Herroelen, W. (2002). *Project Scheduling: A Research Handbook*, Boston, MA: Kluwer Academic Publishers.

Eusuff, M., Lansey, K., and Pasha, F. (2006). Shuffled frog-leaping algorithm: a memetic meta-heuristic for discrete optimization, *Engineering Optimization, 38(2)*, pp. 129–154, https://doi.org/10.1080/03052150500384759.

Golestaneh, R., Jafari, A., Khalilzadeh, M., and Karimi, H. (2013). Minimizing total resource tardiness penalty costs in the resource constrained project scheduling problem with metaheuristic algorithms. *International Journal of Research in Industrial Engineering, 2(3)*, pp. 47–57.

Khalilzadeh, M. (2021). Resource levelling in projects considering different activity execution modes and splitting. *Journal of Engineering, Design and Technology*, Article In Press. https://doi.org/10.1108/JEDT-11-2020-0463.

Khalilzadeh, M., Kianfar, F., Shirzadeh Chaleshtari, A., Shadrokh, S., and Ranjbar, M. (2012). A Modified PSO Algorithm for Minimizing the Total Costs of Resources in MRCPSP. *Mathematical Problems in Engineering, 365697*. https://doi.org/10.1155/2012/365697.

Lova, A., Tormos, P., and Barber, F. (2006). Multi-mode resource constrained project scheduling: scheduling schemes, priority rules and mode selection rules. *Inteligencia Artificial, 10(30)*, pp. 69–86.

Lova, A., Tormos, P., Cervantes, M., and Barber, F. (2009). An efficient hybrid genetic algorithm for scheduling projects with resource constraints and multiple execution modes. *International Journal of Production Economics, 117(2)*, pp. 302–316. https://doi.org/10.1016/j.ijpe.2008.11.002.

Maghsoudlou, H., Afshar-Nadjafi, B., and Niaki, S. T. A. (2020). A framework for preemptive multi-skilled project scheduling problem with time-of-use energy tariffs. *Energy Systems, 96*, pp. 1–28.

Pinedo, M., and Singer, M. (1999). A shifting bottleneck heuristic for minimizing the total weighted tardiness in a job shop. *Naval Research and Logistics, 46*, pp. 1–17.

Vanhoucke, M., and Coelho, J. (2019). Resource-constrained project scheduling with activity splitting and setup times. *Computers & Operations Research, 109*, pp. 230–249.

Vanhoucke, M., Demeulemeester, E., and Herroelen, W. (2001). An exact procedure for the resource-constrained weighted earliness–tardiness project scheduling problem. *Annals of Operations Research, 102*, pp. 179–196. https://doi.org/10.1023/A:1010958200070.

8 Profit Maximization Inventory Control Model with Partial Backlogging and Deterioration on Time-Dependent Demand

Sanjay Sharma and B.B. Verma
Department of Applied Sciences & Humanities,
Ajay Kumar Garg Engineering College, Ghaziabad, India

Anand Tyagi
Department of Mathematics, BKBIET, Pilani, India

CONTENTS

8.1 INTRODUCTION

An effective model for increasing profit is some kind coordination is to conduct the inventory control and manufacturing decision jointly such method of handling inventory is called flexible manufacturing systems. Also, with time inventory depreciates and with it the value of invested capital also decreases. If we look at the multidimensional utilization of flexible manufacturing systems, then we can say that its potential benefits are less waste, low workstations, we can make quick change in handling. Also, it can be seen that because of flexible manufacturing systems facility, we can even work in the direction of making less dies and stamping machinery.

DOI: 10.1201/9781003156291-8

We can also reduce downtime which means that we can take better control over quality; we also reduced labor which means that we have more efficient use of machinery, increased capacity and increased production efficiency.

In 1991, some authors have started to use the concept of production flexibility and they have proved that profit can be maximized by using this concept of production flexibility.

In addition of different applications in the direction of using flexible manufacturing systems, Schweitzer and Seidmann (1991) developed a model based on flexible manufacturing systems also, machine production rate was taken as decision variable which mean that the unit production cost becomes a function of production rate.

Various research papers were presented based on flexible manufacturing systems during 1990 and 1991, in the same direction, Moon et al. (1991) developed inventory models under the assumption of taking common cycle for all the items and published their work in which effect of slowing down was considered where incorporated marketing-based model was developed by Urban (1992).

For stipulation problem of a common cycle for all the items, Gallego (1993) extended the model of Silver (1990) by removing the stipulation.

Economic lot size production was also taking into account to improve production and many research study was published on economic production lot size, Khouja and Mehrez (1994) developed a model based on economic lot size production where Khouja (1995) and Khouja (1996) extended above model by considering imperfect production under the effect of flexible production rate. In the same way, Sana et al. (2007a, b) also make some important changes in the economic production based model and presented an inventory control model under the production system producing items and assumed volume FMS on perfect and imperfect items.

Kumar et al. (2010) worked in the direction of profit maximization techniques and presented production inventory models by considering time-dependent demand and partial backlogging. Sharma et al. (2011) explore the concept of inflation on inventory model and developed an inventory model for low life-based items with inflation, lost sales, stock- and time-dependent demand. For the items where multivariate consumption rate is required to maintain profit, Sharma and Singh (2013) presented an inventory model for decaying items with partial backlogging. For two warehouse-based model, Sharma et al. (2013) developed production policy for the decaying items by assuming different demands for different warehouses under volume flexibility. For the item in which production is assumed as demand dependent, Sharma et al. (2015) developed an economic production quantity model for deteriorating items under permissible shortages.

In the direction of stochastic-based deterioration, Pervin et al. (2018) analyzed inventory control model with shortage under time-dependent demand as well as time-varying holding cost. To improve the profit, a concept of stock-level-dependent production rate, Chen et al. (2019) developed replenishment policy for deteriorating inventory for time-varying and price-dependent demand.

For the low life cycle items, Sharma et al. (2019) presented inventory model with price and time-dependent seasonal demand. Bajegani and Gholamian (2020) developed an optimal inventory control of obsolete products with price-dependent demand.

Palanivel and Suganya (2021) developed partial backlogging inventory model with price and stock-level-dependent demand, time-varying holding cost, and quantity discounts. Shah *et al.* (2021) presented an inventory model with stock-dependent demand under trade-credit policy and fixed lifetime. Akan *et al.* (2021) developed an optimal pricing and inventory strategies for fashion products under time-dependent interest rate and demand.

In this study, we combine all abovementioned factors to make this study more realistic. In this chapter, we have developed the economic production quantity model with time-dependent demand rate and Weibull deterioration. Production rate is decision variable on a volume flexible production policy. Shortages are allowed and partially backlogged. Our model is realistic due to present market situation. The model is developed under the following assumptions and notations.

8.2 ASSUMPTIONS AND NOTATIONS

1. The model is developed for single item.
2. The rate of production is variable.
3. Lead time is zero.
4. Shortages are allowed and partially backlogged.
5. Inventory deteriorates over a time at Weibull distribution rate function.

The following notations have been applied:

K: The tare of production
$D(t) = a + bt$: Linear demand rate, $t \geq 0$
$\theta = \alpha\beta(t - \lambda)^{\beta-1}$: Weibull deterioration rate of the on-hand inventory $\alpha \leq\leq 1, \beta \geq 1$
λ: Lifetime commodity
T: The duration of the production cycle
$\eta(K)$: The production cost per unit item
R: Material cost per unit item
H: Tool/Die cost per unit item
G: Total labor cost for manufacturing item
S_p: Selling price per unit
C_h: Inventory holding cost per unit per unit time
C_s: Shortage cost per unit time
Cs: Setup cost per production run
C_z: Shortage cost per unit per unit time
CI: Lost sale cost
$B = e^{-\delta t}$: Backlogging rate

8.3 FORMULATION AND ANALYSIS OF THE MODEL

Initially, at time $t = 0$ as per graphical representation of the study, we have no item in the inventory. Now, we have assumed that the production starts at time $t = 0$ with the concept of lifetime λ, when the inventory level becomes a certain level thereafter deterioration takes place and after t_1 units of time, it reaches to maximum inventory level. After this, production stopped and at time t_2 the inventory becomes zero due to the combined effect of demand and deterioration. After this, time shortage starts developing and at time t_3 it reaches to maximum shortage level, at this time fresh

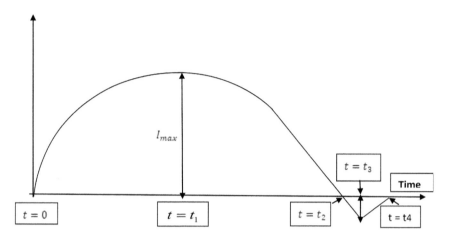

FIGURE 8.1   Graphical representation of the inventory with respect to time.

production starts to clear the backlog by the time t_4. Graphical representation of the model mentioned in Figure 8.1.

The differential equations governing the proposed model are:

$$\frac{dI}{dt} = k - D(t), 0 \leq t \leq \lambda \qquad (8.1)$$

Equation (8.1) showing that between $0 \leq t \leq \lambda$ demand and end production take place

$$\frac{dI}{dt} + \theta(t) = K - D(t), \lambda \leq t \leq t_1 \qquad (8.2)$$

Equation (8.2) showing that, between the time duration $\lambda \leq t \leq t_1$, demand, production and deterioration take place together

$$\frac{dI}{dt} + \theta(t)I(t) = -D(t), t_1 \leq t \leq t_2 \qquad (8.3)$$

Equation (8.3) showing that between $t_1 \leq t \leq t_2$ demand and deterioration take place

$$\frac{dI}{dt} = -BD(t), t_2 \leq t \leq t_3 \qquad (8.4)$$

Equation (8.4) showing that between $t_2 \leq t \leq t_3$ shortages of the item started

$$\frac{dI}{dt} = k - D(t), t_3 \leq t \leq T \qquad (8.5)$$

Equation (8.5) showing that between $t_3 \leq t \leq T$ shortages is satisfying the demand and we have started production.

On putting the value of $D(t)$ and θ, equation becomes,

$$\frac{dI}{dt} = k - (a + bt) \qquad (8.6)$$

$$\frac{dI}{dt} + \alpha\beta(t-\lambda)^{\beta-1}I(t) = k - (a+bt) \tag{8.7}$$

$$\frac{dI}{dt} + \alpha\beta(t-\lambda)^{\beta-1}I(t) = -(a+bt) \tag{8.8}$$

$$\frac{dI}{dt} = -(a+bt)e^{-\delta t} \tag{8.9}$$

$$\frac{dI}{dt} = k - (a+bt) \tag{8.10}$$

Solution of (8.6)–(8.10), respectively, is given as

$$I_1(t) = Kt - b\frac{t^2}{2} \tag{8.11}$$

$$I_2(t) = e^{-\alpha(t-\lambda)^2}$$

$$\times \left[Kt + \frac{\alpha K(t-\lambda)^{\beta+1}}{\beta+1} - at - b\frac{t^2}{2} - \frac{a\alpha(t-\lambda)^{\beta+1}}{\beta+1} - \frac{\alpha b(t-\lambda)^{\beta+1}}{\beta+1} + \frac{\alpha b(t-\lambda)^{\beta+2}}{(\beta+1)(\beta+2)} \right] \tag{8.12}$$

$$I_3(t) = e^{-\alpha(t-\lambda)^2}$$

$$\times \left[a\left(t_2 - t + \frac{\alpha(t_2-\lambda)^{\beta+1}}{\beta+1} - \frac{\alpha(t-\lambda)^{\beta+1}}{\beta+1} \right) + b\left(\frac{t^2_2 - t^2}{2} + \frac{at_2(t_2-\lambda)^{\beta+1}}{\beta+1} - \frac{t\alpha(t-\lambda)^{\beta+1}}{\beta+1} \right) \\ - \frac{\alpha(t_2-\lambda)^{\beta+2}}{(\beta+1)(\beta+2)} + \frac{\alpha(t-\lambda)^{\beta+2}}{(\beta+1)(\beta+2)} \right] \tag{8.13}$$

$$I_4(t) = \left[\frac{ae^{-\delta t}}{\delta} + \frac{bte^{-\delta t}}{\delta} + \frac{be^{-\delta t}}{\delta^2} - \frac{e^{-\delta t_3}}{\delta} - \frac{bt_3e^{-\delta t}}{\delta} - \frac{be^{-\delta t}}{\delta^2} - P \right] \tag{8.14}$$

$$I_5(t) = \left[K(t-T) + a(T-t) + b\left(\frac{T^2-t^2}{2} \right) \right] \tag{8.15}$$

$$\text{Here}, t = t_1, I_3(t) = S_0 \tag{8.16}$$

$$S_0 = e^{-\alpha(t_1-\lambda)^\beta}$$

$$\times \left[a\left(t_2 - t + \frac{\alpha(t_2-\lambda)^{\beta+1}}{\beta+1} - \frac{\alpha(t-\lambda)^{\beta+1}}{\beta+1} \right) \\ + b\left(\frac{t^2_2 - t^2}{2} + \frac{\alpha t_2(t_2-\lambda)^{\beta+1}}{\beta+1} - \frac{\alpha t_1(t_1-\lambda)^{\beta+1}}{\beta+1} + \frac{\alpha(t_2-\lambda)^{\beta+2}}{(\beta+1)(\beta+2)} - \frac{\alpha(t_1-\lambda)^{\beta+2}}{(\beta+1)(\beta+2)} \right) \right] \tag{8.17}$$

From Equations (8.17) and (8.14), we get

$$I_3(t) = S_0 e^{\alpha(t_1-\lambda)^\beta - (t-\lambda)^\beta}$$

$$\left[+e^{-\alpha(t_1-\lambda)^\beta} \begin{bmatrix} a\left(t_1 - t + \dfrac{\alpha(t_1-\lambda)^{\beta+1}}{\beta+1} - \dfrac{\alpha(t-\lambda)^{\beta+1}}{\beta+1} \right) \\[4mm] +b\begin{pmatrix} \dfrac{t_1^2 - t^2}{2} + \dfrac{\alpha t_1(t_1-\lambda)^{\beta+1}}{\beta+1} - \dfrac{\alpha t_1(t_1-\lambda)^{\beta+1}}{\beta+1} \\[4mm] + \dfrac{\alpha(t_1-\lambda)^{\beta+2}}{(\beta+1)(\beta+2)} - \dfrac{\alpha(t-\lambda)^{\beta+2}}{(\beta+1)(\beta+2)} \end{pmatrix} \end{bmatrix} \right] \tag{8.18}$$

Total inventory during the interval $0 \le t \le \lambda, \lambda \le t \le t_1$ and $t_1 \le t \le t_2$

$$Inv_1 = \int_0^\lambda Q_1(t)dt$$

$$= \left[\frac{K\lambda^2}{2} - \frac{a\lambda^2}{2} - \frac{b\lambda^3}{6} \right] \tag{8.19}$$

$$Inv_2 = \int_\lambda^{t_1} Q_2(t)dt$$

$$\left[\frac{K\lambda^2}{2} - \frac{a\lambda^2}{2} - \frac{b\lambda^3}{6} - \frac{Kt_1^2}{2} - \frac{at_1^2}{2} + \frac{bt_1^3}{2} - \frac{\alpha K(t_1-\lambda)^{\beta+2}}{(\beta+1)(\beta+2)} + \frac{2a\alpha(t_1-\lambda)^{\beta+2}}{(\beta+1)(\beta+2)} + \frac{2\alpha b(t_1-\lambda)^{\beta+2}}{(\beta+1)(\beta+2)} \right.$$
$$\left. - \frac{3b\alpha(t_1-\lambda)^{\beta+3}}{(\beta+1)(\beta+2)(\beta+3)} - \frac{\alpha a t_1(t_1-\lambda)^{\beta+1}}{(\beta+1)} - \frac{\alpha b t_1^2(t_1-\lambda)^{\beta+2}}{2(\beta+1)} + \frac{\beta+1}{(\beta+1)} \right] \tag{8.20}$$

$$Inv_3 = \int_{t_1}^{t2} Q_3(t)dt$$

$$\left[\begin{aligned} &\left(S_0 t_2 + \alpha S_0 t_2(t_1-\lambda)^\beta - \frac{\alpha S_0(t_2-\lambda)^{\beta+1}}{\beta+1} - S_0 t_1 - \alpha S_0 t_1(t_1-\lambda)^\beta + \frac{\alpha S_0(t_2-\lambda)^{\beta+1}}{\beta+1} \right) \\[3mm] &+a\begin{pmatrix} t_1 t_2 - \dfrac{t^2_2}{2} + \dfrac{\alpha t_2(t_1-\lambda)^{\beta+1}}{\beta+1} - \dfrac{2\alpha(t_2-\lambda)^{\beta+2}}{(\beta+1)(\beta+2)} \\[3mm] - \dfrac{\alpha t_1(t_2-\lambda)^{\beta+1}}{\beta+1} + \dfrac{\alpha t_2(t_2-\lambda)^{\beta+1}}{\beta+1} - \dfrac{t^2_1}{2} + \dfrac{2\alpha(t_1-\lambda)^{\beta+2}}{(\beta+1)(\beta+2)} - \dfrac{\alpha t_1(t_1-\lambda)^{\beta+2}}{(\beta+1)(\beta+2)} \end{pmatrix} \\[3mm] &+b\begin{pmatrix} \dfrac{t^2_1 t_2}{2} - \dfrac{t^3_2}{6} + \dfrac{\alpha t_1 t_2(t_1-\lambda)^{\beta+1}}{\beta+1} - \dfrac{\alpha t_2(t_1-\lambda)^{\beta+2}}{(\beta+1)(\beta+2)} + \dfrac{\alpha(t_2-\lambda)^{\beta+3}}{(\beta+1)(\beta+2)(\beta+3)} \\[3mm] - \dfrac{\alpha t_1^2(t_1-\lambda)^{\beta+1}}{2(\beta+1)} - \dfrac{\alpha t_1(t_1-\lambda)^{\beta+2}}{(\beta+1)(\beta+2)} \end{pmatrix} \end{aligned} \right] \tag{8.21}$$

Now, total inventory

$$I = Inv_1 + Inv_2 + Inv_3 \tag{8.22}$$

Total no. of deteriorated item is

$$I_D = \theta \left[\int_0^\lambda Q_1(t)dt + \int_\lambda^{t_1} Q_2(t)dt + \int_{t_1}^{t_2} Q_3(t)dt \right] \tag{8.23}$$

$$= \theta I$$

Total shortage in inventory is given by

$$S = \int_{t_2}^{t_3} -I_4(t)dt$$

$$= - \left[\begin{array}{l} \dfrac{ae^{-\delta t_3}}{-\delta^2} + \dfrac{2bt_3 e^{-\delta t_3}}{\delta^2} - \dfrac{2be^{-\delta t_3}}{\delta^3} - \dfrac{at_3 e^{-\delta t_3}}{\delta} - \dfrac{bt^2{}_3 e^{-\delta t_3}}{2\delta} - Pt_3 + Pt_2 + \dfrac{ae^{-\delta t_2}}{\delta^2} \\[3mm] + \dfrac{bt_2 e^{-\delta t_2}}{\delta^2} + \dfrac{2be^{-\delta t_2}}{\delta^3} + \dfrac{at_2 e^{-\delta t_3}}{\delta} + \dfrac{bt_2 t_3 e^{-\delta t_3}}{2\delta} + \dfrac{2t_2 e^{-\delta t_3}}{\delta^2} \end{array} \right] \tag{8.24}$$

Lost sale in inventory is given by

$$L.S = \int_{t_3}^{T} (1 - e^{-\delta t})dt$$

$$= \left[aT + \dfrac{bT^2}{2} + \dfrac{bTe^{-\delta T}}{\delta^2} + \dfrac{be^{-\delta T}}{\delta^2} - at_3 - \dfrac{bt^2{}_3}{2} - \dfrac{ae^{-\delta t_3}}{\delta} - \dfrac{be^{-\delta t_3}}{\delta^2} \right] \tag{8.25}$$

The average profit during time interval $(0,T)$

$$\prod(K, t_2, T) = \dfrac{1}{T}(Kt_2 - I_D)S_p$$

$$- \dfrac{1}{T}(\{C_s + C_h(Inv_1 + Inv_2 + Inv_3) + C_2 S - C_1 L.S.) - \dfrac{1}{T}(RK + G + HK^2)t_2\} \tag{8.26}$$

In this manner, we obtain the total profit function. We can solve the said equation by using the profit maximization technique with the help of some suitable computational mathematical software.

8.4 SENSITIVITY ANALYSIS

Based on the values used in above example in the model, authors have examined the sensitivity analysis by changing some parameters one at a time and keeping the rest fixed.

Authors have calculated the sensitivity analysis based on different parameters. The outcome of the result is compared.

8.5 OBSERVATIONS

In this study, authors have calculated the sensitivity analysis based on the parameters used in the study. Authors have made changes in the parameters by −50%, −25%, 0%, 25%, 50%. Some important inferences drawn from Table 8.1 and Figures 8.2–8.5 are as follows:

  i. Table 8.1 shows that as the value of a goes from −50% to +50%, the value of t_1, T, I_{max}, and total cost also increases.
 ii. Table 8.1 shows that as the value of a goes from −50% to +50%, the value of T and I_{max} decreases while the value of t_1 and total cost increases rapidly.
iii. Table 8.1 shows that as the value of b goes from −50% to +50%, the value of T total cost increases while the value of t_1, T, and I_{max} decrease rapidly.
 iv. Table 8.1 shows that as the value of p goes from −50% to +50%, the value of T total cost increases while the value of t_1, T, and I_{max} decreases

TABLE 8.1
Sensitivity Analysis w.r.t. Various Parameters

Parameter	%	Changed Value	t_1	T	I_{max}	TC
H	+50	75	1.34725	2.69451	169.0449	359.918
	+25	62.5	1.22741	2.40402	130.22971	354.026
	0	50	1.10003	1.20006	114.77403	349.788
	−25	37.5	0.97816	1.98821	71.21354	341.498
	−50	25	0.87699	1.68724	32.21354	333.805
G	+50	30	2.43291	1.09105	19.33721	489.125
	+25	25	1.22886	1.10554	53.97692	431.226
	0	20	1.10003	2.20006	114.77403	349.788
	−25	15	0.92214	2.31269	148.29284	266.394
	−50	10	0.78764	2.39991	187.76542	200.947
Ch	+50	1.8	1.09371	2.18739	90.53786	354.985
	+25	1.4	1.09685	2.19369	102.73451	352.063
	0	1.2	1.10003	2.20006	114.77403	339.788
	−25	0.8	1.10324	2.20647	123.95031	348.072
	−50	0.6	1.12009	2.22018	149.24731	346.857
Cz	+50	56	0.53185	1.59475	107.98073	361.688
	+25	50	0.86818	1.82913	110.76742	357.296
	0	40	1.10003	2.20006	114.77403	349.788
	−25	30	1.33422	2.62627	143.91819	335.358
	−50	20	1.67151	2.91326	201.21387	317.963

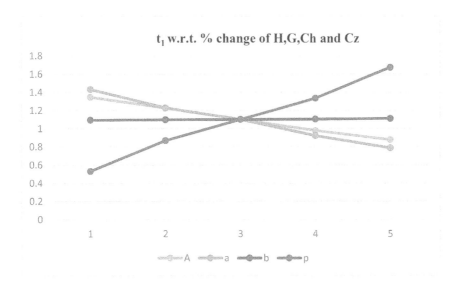

FIGURE 8.2 t_1 v/s change in parameters.

8.6   CONCLUSION

Current study employs an approach to handle deterioration which is assumed to follow a three parameter Weibull distribution which make this study more valuable also, we have considered shortages which is partially backlogged. The utilization of a three parameter Weibull distribution makes the scope of the application broader. In this study, we have implemented a new technique of flexibility in the machine production

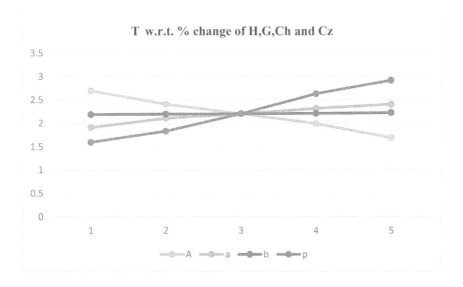

FIGURE 8.3 T v/s change in parameters.

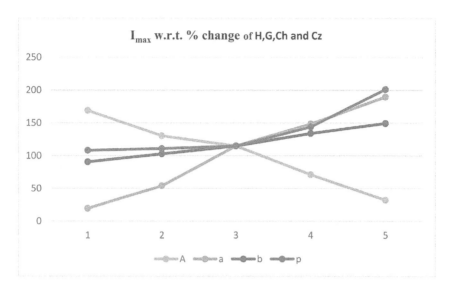

FIGURE 8.4 I_{max} v/s change in parameters.

rate. The purpose of this chapter is to incorporate the concept of flexibility in the machine production rate. Unit production cost is linked to the variable production rate.

Past research in the area of production inventory modeling has yielded a variety of inventory models, which suits several situations. Yet there is a wide scope as well as need of volume flexibility in the corresponding models to make it more realistic due to present business situation. When the manufacturer companies have greater control over the production, they are able to manipulate it so that the cost is minimized and

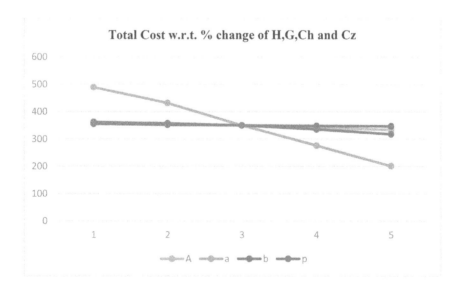

FIGURE 8.5   Total cost v/s change in parameters.

profit is maximized. This strategy is essential options that small and large firms use to gain competitive advantage. The model is developed under time-depended deterioration and shortages are not allowed. The net profit function has been maximized for optimality.

It is also found that considering different demands on different stages of model is more economical to maximize the profit. Further, this model can be expanded for further utilization by adding or involving a few more parameters like permissible delay in the parameter and more dependence of demand and productions.

8.6.1 FUTURE SCOPE OF THE STUDY

The study may be more expanded by making some more changes in the mail. Parameters of the study like this model can also be developed under a shortage, trade-credits may also be applied in this study.

REFERENCES

Akan, M., Albey, E. and Güler, M.G. (2021), "Optimal pricing and inventory strategies for fashion products under time-dependent interest rate and demand." Computers & Industrial Engineering, 154, 107149.

Bajegani, H.Z. and Gholamian, M.R. (2020), "Optimal inventory control of obsolete products with price-dependent demand." Journal of Engineering Research, 8(4), pp. 169–184.

Chen, L., Chen, X., Keblis, M.F. and Li, G. (2019), "Optimal pricing and replenishment policy for deteriorating inventory under stock-level-dependent, time-varying and price-dependent demand." Computers & Industrial Engineering, 135, 1294–1299.

Gallego, G. (1993), "Reduced production rates in the economic lot scheduling problem." International Journal of Production Research, 31(5), 1035–1046.

Khouja, M. (1995), "The economic production lot size model under volume flexibility." Computer and operation Research, 22(5), 515–523.

Khouja, M. (1996), "The scheduling of economic lot size on volume flexibility production system". International Journal of Production Economics, 48(1) 73–86.

Khouja, M. and Mehrez, A. (1994), "An economic production lot size model with imperfect quality and variable production rate." Journal of Operational Research Society, 45(12), 1405–1417.

Kumar et al. (2010), "Profit maximization production inventory models with time dependent demand and partial backlogging." International Journal of Operations Research and Optimization, 2(3), 367–375.

Moon, I., Gallego, G. and Simchi-Levi, D. (1991), "Controllable production rate in a family production context." International Journal of Production Research, 29, 2459–2470.

Palanivel, M. and Suganya, M. (2021), "Partial backlogging inventory model with price and stock level dependent demand, time varying holding cost and quantity discounts." Journal of Management Analytics, 1–28.

Pervin, M., Roy, S.K. and Weber, G.W. (2018), "Analysis of inventory control model with shortage under time-dependent demand and time-varying holding cost including stochastic deterioration." Annals of Operations Research, 260(1), 437–460.

Sana et al. (2007a), "On a volume flexible inventory model for items with an imperfect production system." International Journal of Operational Research, 2(1), 64–80.

Sana et al. (2007b), "An Imperfect production process in a volume flexible inventory model." International Journal of Production Economics, 105, 548–559.

Schweitzer, P.J. and Seidmann, A. (1991), "Optimizing processing rate for flexible manufacturing systems." Management Science, 37, 454–466.

Shah, N.H., Rabari, K. and Patel, E. (2021), "An inventory model with stock-dependent demand under trade-credit policy and fixed life-time." International Journal of Procurement Management, 14(2), 264–275.

Sharma, P., Sharma, A. and Jain, S. (2019), "Inventory model for deteriorating items with price and time-dependent seasonal demand." International Journal of Procurement Management, 12(4), 363–375.

Sharma, S. and Singh, S.R. (2013), "An inventory model for decaying items, considering multivariate consumption rate with partial backlogging." Indian Journal of Science and Technology, 6(7), 4870–4880.

Sharma, S., Singh, S.R. and Dem, H. (2013), "Two-warehouse production policy for different demands under volume flexibility." International Journal of Industrial Engineering Computations Science, 4, 599–609.

Sharma, S., Singh, S.R. and Ram, M. (2015), "An EPQ model for deteriorating items with price sensitive demand and shortages in which production is demand dependent." International Journal of Mathematics in Operational Research, 23(2), 245–255.

Silver (1990), "Chromium reduction in Pseudomonas putida." Applied and Environmental Microbiology, 56(7), 2268–2270.

Urban, T.L. (1992), "Deterministic inventory models incorporating marketing decision." Computers and Industrial Engineering, 22(1), 85–93.

9 On Joint Importance Measures for Multistate System's Reliability

V.M. Chacko
Department of Statistics, St. Thomas College (Autonomous),
Thrissur, University of Calicut, Kerala, India

Nomenclature: The following nomenclatures are used throughout the chapter.

Abbreviation	Name
MSS	*Multistate system*
OPM	*Output performance measure*
IM	*Importance measure*
RAW	*Reliability Achievement Worth*
RRW	*Reliability Reduction Worth*
FV	*Fussell-Vesely*
UGF	*Universal generating function*
JBRAW	*Joint Birnbaum-Reliability Achievement Worth*
JBRRW	*Joint Birnbaum-Reliability Reduction Worth*
JBRFV	*Joint Birnbaum-Reliability Fussell-Vesely*
JBrAW	*Joint Birnbaum Risk Achievement Worth*
JBrRW	*Joint Birnbaum Risk Reduction Worth*
JBrFV	*Joint Birnbaum Risk Fussell-Vesely*
MJBRAW	*Multistate Joint Birnbaum-Reliability Achievement Worth*
MJBRRW	*Multistate Joint Birnbaum-Reliability Reduction Worth*
MJBRFV	*Multistate Joint Birnbaum-Reliability Fussell-Vesely*
MJBOPMAW	*Multistate Joint Birnbaum Output Performance Measure Achievement Worth*
MJBOPMRW	*Multistate Joint Birnbaum Output Performance Measure Reduction Worth*
MJBOPMFV	*Multistate Joint Birnbaum Output Performance Measure Fussell-Vesely*
MJBrAW	*Multistate Joint Birnbaum-Risk Achievement Worth*
MJBrRW	*Multistate Joint Birnbaum- Risk Reduction Worth*
MJBrFV	*Multistate Joint Birnbaum Risk Fussell-Vesely measure*
MSEOPM	*Multistate Expected Output Performance Measure*
p_i	*Steady-state probability of the MSS state i*
$E(W)$	*Steady-state expected output performance measure*
$E_s(\varphi(X))$	*Steady-state expected system state*
$A(t)$	*Availability for constant demand at time t*
$A(D_k)$	*Steady-state availability for constant demand D_k*
$F(t)$	*Risk metric at time t*
F	*Steady-state risk metric*

DOI: 10.1201/9781003156291-9

CONTENTS

9.1   INTRODUCTION

In binary and multistate reliability engineering, the problem of identification of most important component or more important group of components with regard to the contribution of change in performance measures like reliability, risk and expected system performance measures is to be addressed for system performance measure improvement activities. Measures of importance are used to identify most important components or group of components in a system with regard to the various performance measures. To order the components of the system in reliability engineering, there are several importance measures (IMs) available in literature (Barlow and Proschan, 1975). See Barlow and Wu (1978) and Bueno (1989) for fundamental details of IMs of multistate systems (MSSs). IMs help the system designers in improving system performance (Zio et al., 2007). Joint IMs are used to identify the group of components, the improvement of which results in highest improvement in system performance. Interaction importance of groups of components, with respect to various output performance measure (OPM)s, reliability, expected output performance, risk etc., is more helpful to the designers, engineers and managers (Wu and Chan, 2003; Wu, 2005; Chacko and Manoharan, 2008, 2011; Chacko, 2020, 2021).

In risk-informed applications of the nuclear industry, analysis using IMs is crucial (see Vasseur and Llory, 1999; Cheok, Parry, and Sherry, 1998; Van der Borst and Shoonakker, 2001; Borgonovo and Apostolakis, 2001). IMs are also often used for ranking the components with respect to their contribution to the considered metric of risk-unreliability or unavailability.

The joint IMs of three components for MSS with respect to various OPMs like reliability, expected output performance and risk to the existing measures of importance are unexplored one. In this chapter, as three components of binary and MSS, Joint Birnbaum-Reliability Achievement Worth (*JBRAW*), Joint

Birnbaum-Reliability Reduction Worth (*JBRRW*) and Joint Birnbaum-Reliability Fussell-Vesely (*JBRFV*) IMs are introduced and generalized to various OPMs like expected OPM and availability. Joint Birnbaum Risk Achievement Worth (*JBrAW*), joint Birnbaum-Risk Reduction Worth (*JBrRW*) and Joint Birnbaum Risk Fussell-Vesely (*JBrFV*) IMs are also introduced for three components. *JBRAW* measures the potential change in *RAW* of one component when interaction effect of other two components changes from higher level to lower level, and *JBRRW* measures the potential change in *RRW* of one component when interaction effect of other two components change from higher level to lower level and *JBRFV* measures the potential change in Fussell-Vesely (*FV*) measure of one component when interaction effect of other two components change from higher level to lower level. A steady-state performance level restriction method is given for obtaining the proposed measures. The information derived by these joint IMs allows the analyst to judge how important it is to keep the performance of three given components, based on their interaction effect for system OPM improvement. Let OPM_i^α be the system OPM obtained when component i is constrained to performance below α, while the rest of components of the MSS are not constrained. Similarly, denote by $OPM_i^{>\alpha}$ the system OPM in which component i is constrained to performances above α. Also similarly let $OPM_{i,j}^{>\alpha,>\beta}$, $OPM_{i,j}^{\leq\alpha,>\beta}$, $OPM_{(i,j)}^{(>\alpha,\leq\beta)}$, $OPM_{(i,j)}^{(\leq\alpha,\leq\beta)}$, $OPM_{i,j,q}^{>\alpha,>\beta,>\delta}$, $OPM_{i,j,q}^{\leq\alpha,>\beta,>\delta}$, $OPM_{i,j,q}^{>\alpha,\leq\beta,>\delta}$, $OPM_{i,j,q}^{>\alpha,>\beta,\leq\delta}$, $OPM_{i,j,q}^{\leq\alpha,\leq\beta,\leq\delta}$, $OPM_{i,j,q}^{>\alpha,\leq\beta,\leq\delta}$, $OPM_{i,j,q}^{\leq\alpha,>\beta,\leq\delta}$, $OPM_{i,j,q}^{\leq\alpha,\leq\beta,>\delta}$, and $OPM_{i,j,q}^{\leq\alpha,\leq\beta,\leq\delta}$, be the OPMs when the components i, j and q are restricted in their performance with respect to performance thresholds α, β and δ, respectively. In power generation, oil transportation systems etc., the performance measure of series system will be sum performance measure of components instead of minimum. In such complex systems, Universal generating function (UGF) method is commonly used to evaluate system performance.

In Section 9.2, the performance measures of the MSS are reviewed. In Section 9.3, new joint IMs for three components of the MSS are introduced. In Section 9.4, an approach of element performance restriction for the evaluation of performance measures is given. Also, a technique for joint IMs evaluation based on the UGF method is given. A numerical example is provided in Section 9.5. Results and discussion are given in Section 9.6. Conclusion is given in Section 9.7.

9.2 PERFORMANCE MEASURES OF A MULTISTATE SYSTEM

Let the structure function of an MSS at time t be $\varphi(X(t))$, $\varphi(X(t)) \in \{0,1,2,...,M\}$, where $X(t) = (X_1(t), X_2(t),..., X_n(t))$, $X_i(t) \in \{0,1,2,...,M_i\}$, $i = 1,2,...,n$ are the vector of component states and $M = \max_{1 \leq i \leq n}\{M_i\}$. The output performance of the MSS at time t is denoted by $W(t) \in \{w_i, i = 0,1,...,M\}$ where w_i is the performance corresponding to the system state $\varphi(X(t)) = i$.

Steady-state probability distribution of the MSS state is:

$$p_i = \lim_{t\to\infty} \Pr\{W(t) = w_i\} = \lim_{t\to\infty} \Pr\{\phi(X(t)) = i\}, \quad 0 \leq i \leq M.$$

Let $\mathbf{w} = \{w_i, 0 \leq i \leq M\}$ be the output performance of system and $\mathbf{p} = \{p_i, 0 \leq i \leq M\}$ be the corresponding probabilities. Steady-state expected value is

$$E(W) = \sum_{i=0}^{M} p_i w_i \tag{9.1}$$

Similarly, steady-state expected system state is

$$E_s(\phi(X)) = \sum_{i-0}^{M} i p_i \tag{9.2}$$

MSS availability for constant demand D_k, to state k of the system is

$$A(t) = \Pr\{W(t) \geq D_k\} = \Pr\{\phi(t) \geq k\} \tag{9.3}$$

From Equations (9.1) and (9.3), the stationary availability is

$$A(D_k) = \sum_{i=0}^{M} p_i 1(w_i - D_k). \tag{9.4}$$

The metric of risk, F, at time t, is

$$F(t) = \Pr\{W(t) \leq D_k\} = \Pr\{\phi(X(t) \leq k\} \tag{9.5}$$

which represents the unreliability or unavailability. Steady-state risk metric is

$$F = \lim_{x \to \infty} \Pr\{W(t) \leq D_k\} \tag{9.6}$$

In the following section, new joint IMs, joint IM of one component and interaction performance of other two components, are proposed.

For ith multistate component i and performance threshold α, let $k_{i\alpha}$ be the state in the ordered set of states of component i with performance $x_{ik_{i\alpha}}$ equal or immediately below α, i.e., $x_{ik_{i\alpha}} \leq \alpha < x_{ik_{i\alpha}+1}$. Similarly, one can restrict the performance of other components with pre-specified thresholds.

9.3 NEW JOINT IMPORTANCE MEASURES FOR THREE COMPONENTS IN THE MSS

In this section, three joint IMs are proposed. Suppose now the components are statistically independent and reliabilities are available.

9.3.1 JOINT BIRNBAUM-RELIABILITY ACHIEVEMENT WORTH (*JBRAW*)

The *JBRAW* measure quantifies the maximum possible deviation in *RAW* due to a particular component i when interaction effect of component j and q switches from high level to low level, while both components j and q move from high level to low level, simultaneously or individually, $i \neq j \neq q$, i, j, $q \in \{1, 2, \ldots, n\}$, n is the number of components in the system.

JBRAW of two binary components i and j when third component q is in perfect functioning state $X_q(t) = 1$, is

$$JBRAW_{ijq1} = \frac{\Pr\{\phi(X(t)) = 1, X_j(t) = 1, X_q(t) = 1 \mid X_i(t) = 1\}}{\Pr\{\phi(X(t)) = 1, X_j(t) = 1, X_q(t) = 1\}}$$

$$- \frac{\Pr\{\phi(X(t)) = 1, X_j(t) = 0, X_q(t) = 1 \mid X_i(t) = 1\}}{\Pr\{\phi(X(t)) = 1, X_j(t) = 0, X_q(t) = 1\}}$$

where $i, j, q \in \{1, 2, \ldots, n\}$. Similarly we can define $JBRAW_{ijq0}$ when $X_q(t) = 0$.

Thus, *JBRAW* of three binary components i, j and q can be defined, in Birnbaum sense, as

$$JBRAW_{ijq} = JBRAW_{ijq1} - JBRAW_{ijq0}, \text{where } i, j, q \in \{1, 2, \ldots, n\}.$$

For a constant demand D_k, multistate Joint Birnbaum-Reliability Achievement Worth (*MJBRAW*) of component i and component j, when third component q is in state $X_q(t) > x_{qk_{q\delta}}$, is defined as,

$$MJBRAW_{ijq>} = \frac{\Pr\{\phi(X(t)) \geq k, X_j(t) > x_{jk_{j\beta}} 1, X_q(t) > qk_{q\delta} \mid X_i(t) > x_{ik_{i\alpha}}\}}{\Pr\{\phi(X(t)) \geq k, X_j(t) > x_{jk_{j\beta}}, X_q(t) > x_{qk_{q\delta}}\}}$$

$$- \frac{\Pr\{\phi(X(t)) \geq k, X_j(t) \leq x_{jk_{j\beta}} 1, X_q(t) > qk_{q\delta} \mid X_i(t) > x_{ik_{i\alpha}}\}}{\Pr\{\phi(X(t)) \geq k, X_j(t) \leq x_{jk_{j\beta}}, X_q(t) > x_{qk_{q\delta}}\}}$$

where α is the performance threshold and $x_{ik_{i\alpha}}$ is the performance of component i in state $k_{i\alpha}$, β is the performance threshold and $x_{jk_{j\beta}}$ is the performance of component j in the state $k_{j\beta}$, δ is the performance threshold and $x_{qk_{q\delta}}$ is the performance of component q in the state $k_{q\delta}$, $i, j, q=1,2,\ldots,n$. Similarly *MJBRAW* of components i and j is defined, when third component q is in state $X_q(t) \leq x_{qk_{q\delta}}$,

$$MJBRAW_{ijq\leq} = \frac{\Pr\{\phi(X(t)) \geq k, X_j(t) > x_{jk_{j\beta}} 1, X_q(t) \leq qk_{q\delta} \mid X_i(t) > x_{ik_{i\alpha}}\}}{\Pr\{\phi(X(t)) \geq k, X_j(t) > x_{jk_{j\beta}}, X_q(t) \leq x_{qk_{q\delta}}\}}$$

$$- \frac{\Pr\{\phi(X(t)) \geq k, X_j(t) \leq x_{jk_{j\beta}} 1, X_q(t) \leq qk_{q\delta} \mid X_i(t) > x_{ik_{i\alpha}}\}}{\Pr\{\phi(X(t)) \geq k, X_j(t) \leq x_{jk_{j\beta}}, X_q(t) \leq x_{qk_{q\delta}}\}}$$

where $i, j, q \in \{1, 2, \ldots, n\}$.

Thus, *MJBRAW* of three components i, j and q can be defined as,

$$MJBRAW_{ijq} = MJBRAW_{ijq>} - MJBRAW_{ijq\leq}, \quad i, j, q = 1, 2, \ldots, n. \tag{9.7}$$

MJBRAW measures the joint effect of reliability achievement worth of component i and the interaction effect of components j and q, measured in Birnbaum sense.

9.3.2 Joint Birnbaum-Reliability Reduction Worth (*JBRRW*)

The *JBRRW* is an index measuring the maximum change in *RRW* of a particular component i when interaction effect of component j and q switches from high level to low level, while both components j and q move from high level to low level simultaneously or individually, $i \neq j \neq q$, i, j, $q \in \{1,2,...,n\}$, n is the number of components in the system.

JBRRW of two binary components i and j when third component q is in perfect functioning state $X_q(t) = 1$ is defined as

$$JBRRW_{ijq1} = \frac{\Pr\left\{\phi(X(t)) = 1, X_j(t) = 1, X_q(t) = 1\right\}}{\Pr\{\phi(X(t)) = 1, X_j(t) = 1, X_q(t) = 1 \mid X_i(t) = 0\}}$$

$$- \frac{\Pr\left\{\phi(X(t)) = 1, X_j(t) = 0, X_q(t) = 1\right\}}{\Pr\{\phi(X(t)) = 1, X_j(t) = 0, X_q(t) = 1 \mid X_i(t) = 0\}}$$

where i, j, $q \in \{1,2,...,n\}$. Similarly we can define, $JBRAW_{ijq0}$ when $X_q(t) = 0$.
JBRRW of three binary components i, j and q can be defined as,

$$JBRRW_{ijq} = JBRRW_{ijq_1} - JBRRW_{ijq_0} \text{ where } i,\ j,\ q \in \{1,2,...,n\}.$$

For a constant demand D_k, Multistate Joint Birnbaum-Reliability Reduction Worth (*MJBRRW*) of components i and j, when third component q is in state $X_q(t) > x_{qk_{q\delta}}$, is defined as,

$$MJBRRW_{ijq>} = \frac{\Pr\{\phi(X(t)) \geq k, X_j(t) > x_{jk_{j\beta}}, X_q(t) > x_{qk_{q\delta}}\}}{\Pr\{\phi(X(t)) \geq k, X_j(t) > x_{jk_{j\beta}}, X_q(t) > qk_{q\delta} \mid X_i(t) \leq x_{ik_{i\alpha}}\}}$$

$$- \frac{\Pr\{\phi(X(t)) \geq k, X_j(t) \leq x_{jk_{j\beta}}, X_q(t) > x_{qk_{q\delta}}\}}{\Pr\{\phi(X(t)) \geq k, X_j(t) \leq x_{jk_{j\beta}} 1, X_q(t) > qk_{q\delta} \mid X_i(t) \leq x_{ik_{i\alpha}}\}}$$

where α is the performance threshold and $x_{ik_{i\alpha}}$ is the performance of i in state $k_{i\alpha}$, β is the performance threshold and $x_{jk_{j\beta}}$ is the performance of component j in the state $k_{j\beta}$, δ is the performance threshold and $x_{qk_{q\delta}}$ is the performance of component q in the state $k_{q\delta}$, $i, j, q = 1,2,...,n$. Similarly define is the *MJBRRW* of components i and j, when third component q is in state $X_q(t) \leq x_{qk_{q\delta}}$,

$$MJBRRW_{ijq\leq} = \frac{\Pr\{\phi(X(t)) \geq k, X_j(t) > x_{jk_{j\beta}}, X_q(t) \leq x_{qk_{q\delta}}\}}{\Pr\{\phi(X(t)) \geq k, X_j(t) > x_{jk_{j\beta}}, X_q(t) \leq qk_{q\delta} \mid X_i(t) \leq x_{ik_{i\alpha}}\}}$$

$$- \frac{\Pr\left\{\phi(X(t)) \geq k, X_j(t) \leq x_{jk_{j\beta}}, X_q(t) \leq x_{qk_{q\delta}}\right\}}{\Pr\{\phi(X(t)) \geq k, X_j(t) \leq x_{jk_{j\beta}} 1, X_q(t) \leq qk_{q\delta} \mid X_i(t) \leq x_{ik_{i\alpha}}\}}$$

Thus, *MJBRRW* of three components i, j and q can be defined as,

$$MJBRRW_{ijq} = MJBRRW_{ijq>} - MJBRRW_{ijq\leq}, \quad i,j,q = 1,2,...,n.$$

MJBRRW measures the joint effect of reliability reduction worth of component i and the interaction effect of components j and q, measured in Birnbaum sense.

9.3.3 JOINT BIRNBAUM-RELIABILITY FUSSELL-VESELY (*JBRFV*) MEASURE

The *JBRFV* IM quantifies the maximum change in reliability FV importance of a particular component i when interaction effect of component j and q switches from high level to low level, while both components j and q move from high level to low level simultaneously or individually, $i \neq j \neq q$, i, j, $q \in \{1,2,...,n\}$, n is the number of components in the system.

JBRFV of two binary components i and j when third component q is in perfect functioning state $X_q(t) = 1$ is defined as

$$JBRFV_{ijq1} = \frac{\Pr\left(\varphi\left(X(t)\right)=1, X_j(t)=1, X_q(t)=1\right) - P\left(\varphi\left(X(t)\right)=1, X_j(t)=1, X_q(t)=1 \mid X_i(t)=0\right)}{\Pr\left(\varphi\left(X(t)\right)=1, X_j(t)=1, X_q(t)=1\right)}$$

$$-\frac{\Pr\left(\varphi\left(X(t)\right)=1, X_j(t)=0, X_q(t)=1\right) - \Pr\left(\varphi\left(X(t)\right)=1, X_j(t)=0, X_q(t)=1 \mid X_i(t)=0\right)}{\Pr\left(\varphi\left(X(t)\right)=1, X_j(t)=0, X_q(t)=1\right)}$$

JBRFV of two binary components i and j when third component q is in failed state $X_q(t) = 0$, for i, j, $q \in \{1,2,...,n\}$, is defined as

$$JBRFV_{ijq0} = \frac{\Pr\left(\varphi\left(X(t)\right)=1, X_j(t)=1, X_q(t)=0\right) - P\left(\varphi\left(X(t)\right)=1, X_j(t)=1, X_q(t)=0 \mid X_i(t)=0\right)}{\Pr\left(\varphi\left(X(t)\right)=1, X_j(t)=1, X_q(t)=0\right)}$$

$$-\frac{\Pr\left(\varphi\left(X(t)\right)=1, X_j(t)=0, X_q(t)=0\right) - \Pr\left(\varphi\left(X(t)\right)=1, X_j(t)=0, X_q(t)=0 \mid X_i(t)=0\right)}{\Pr\left(\varphi\left(X(t)\right)=1, X_j(t)=0, X_q(t)=0\right)}$$

where i, j, $q \in \{1,2,...,n\}$.

Thus, *JBRRW* of three binary components i, j and q can be defined as,

$$JBRFV_{ijq} = JBRFV_{ijq1} - JBRFV_{ijq0}, \text{ where } i, j, q \in \{1,2,...,n\}.$$

For a constant demand D_k, Multistate Joint Birnbaum-Reliability Fussell-Vesely (*MJBRFV*) of components i and j, when third component q is in state $X_q(t) > x_{qk_q\delta}$, is defined as,

$MJBRFV_{ijq>}$

$$=\frac{\Pr\left(\varphi(X(t)) \geq k, X_j(t) > x_{jk_j\beta}, X_q(t) > x_{qk_q\delta}\right) - P\left(\varphi(X(t)) \geq k, X_j(t) > x_{jk_j\beta}, X_q(t) > x_{qk_q\delta} \mid X_i(t) \leq x_{ik_i\alpha}\right)}{\Pr\left(\varphi(X(t)) \geq k, X_j(t) > x_{ik_j\beta}, X_q(t) > x_{qk_q\delta}\right)}$$

$$-\frac{\Pr\left(\varphi(X(t)) \geq k, X_j(t) \leq x_{jk_j\beta}, X_q(t) > x_{qk_q\delta}\right) - \Pr\left(\varphi(X(t)) \geq k, X_j(t) \leq x_{jk_j\beta}, X_q(t) > x_{qk_q\delta} X_i(t) \leq x_{ik_i\alpha}\right)}{\Pr(\varphi(X(t)) \geq k, X_j(t) \leq x_{jk_j\beta}, X_q(t) > x_{qk_q\delta})}$$

where α is the performance threshold and $x_{ik_{i\alpha}}$ performance of component i in state $k_{i\alpha}$, β is the performance threshold and $x_{jk_{j\beta}}$ is the performance of component j in the state $k_{j\beta}$, δ is the performance threshold and $x_{qk_{q\delta}}$ is the performance of component q in the state $k_{q\delta}$, i, j, $q=1,2,...,n$. Similarly defined is the *MJBRFV* of components i and j, when third component q is in state $X_q(t) \leq x_{qk_{q\delta}}$ as,

$MJBRFV_{ijq\leq}$

$$= \frac{\Pr\left(\varphi(X(t)) \geq k, X_j(t) \leq x_{jk_{j\beta}}, X_q(t) \leq x_{qk_{q\delta}}\right) - P\left(\varphi(X(t)) \geq k, X_j(t) > x_{jk_{j\beta}}, X_q(t) \leq x_{qk_{q\delta}} \big| X_i(t) \leq x_{ik_{i\alpha}}\right)}{\Pr\left(\varphi(X(t)) \geq k, X_j(t) > x_{jk_{j\beta}}, X_q(t) \leq x_{qk_{q\delta}}\right)}$$

$$\frac{\Pr\left(\varphi(X(t)) \geq k, X_j(t) \leq x_{jk_{j\beta}}, X_q(t) \leq x_{qk_{q\delta}}\right) - \Pr(\varphi(X(t)) \geq k, X_j(t) \leq x_{jk_{j\beta}}, X_q(t) \leq x_{qk_{q\delta}} X_i(t) \leq x_{ik_{i\alpha}}}{\Pr(\varphi(X(t)) \geq k, X_j(t) \leq x_{jk_{j\beta}}, X_q(t) \leq x_{qk_{q\delta}})}$$

Thus, *MJBRFV* of three components i, j and q can be defined as,

$$MJBRFV_{ijq} = MJBRFV_{ijq>} - MJBRFV_{ijq\leq}, i, j, q = 1,2,...,n.$$

MJBRFV measures the joint effect of reliability *FV* of component i and the interaction of components j and q, measured in Birnbaum sense. When the performance measure changes, the proposed measures need to redefine. Define Multistate Joint Birnbaum Output Performance Measure Achievement Worth (*MJBOPMAW*), Multistate Joint Birnbaum Output Performance Measure Reduction Worth (*MJBOPMRW*) and Multistate Joint Birnbaum Output Performance Measure Fussell-Vesely (*MJBOPMFV*) as below.

$$\text{Let } MJBOPMAW_{ijq<} = \frac{OPM_{i,j}^{>\alpha,>\beta,\leq\delta}}{OPM_{j,q}^{>\beta,\leq\delta}} - \frac{OPM_{i,j}^{>\alpha,\leq\beta,\leq\delta}}{OPM_{j,q}^{\leq\beta,\leq\delta}} \text{ and } MJBOPMAW_{ijq>}$$

$$= \frac{OPM_{i,j}^{>\alpha,>\beta,>\delta}}{OPM_{j,q}^{>\beta,>\delta}} - \frac{OPM_{i,j}^{>\alpha,\leq\beta,>\delta}}{OPM_{j,q}^{\leq\beta,>\delta}}$$

$i, j, q = 1,2,...,n$. Thus, *MJBOPMAW* of three components i, j and q can be defined as,

$$MJBOPMAW_{ijq} = MJBOPMAW_{ij1} - MJBOPMAW_{ij0}, \ i,j,q = 1,2,...,n.$$

$$\text{Let } MJBOPMRW_{ijq\leq} = \frac{OPM_{j,q}^{>\beta,\ \leq\delta}}{OPM_{i,j,q}^{\leq\alpha,>\beta,\leq\delta}} - \frac{OPM_{j,q}^{\leq\beta,\leq\delta}}{OPM_{i,j,q}^{\leq\alpha,\leq\beta,\leq\delta}} \text{ and }$$

$$MJBOPMRW_{ijq>} = \frac{OPM_{j,q}^{>\beta,\ >\delta}}{OPM_{i,j,q}^{\leq\alpha,>\beta,>\delta}} - \frac{OPM_{j,q}^{\leq\beta,>\delta}}{OPM_{i,j,q}^{\leq\alpha,\leq\beta,>\delta}}, i, j, q = 1,2,...,n.$$

Thus, *MJBOPMRW* of three components i, j and q can be defined as,

$$MJBOPMRW_{ijq} = MJBOPMRW_{ij>} - MJBOPMRW_{ij\leq} \ i, j, q = 1,2,...,n.$$

Let $\quad MJBOPMFV_{ijq\leq} = \dfrac{OPM_{j,q}^{>\beta,\,\leq\delta} - OPM_{i,j,q}^{\leq\alpha,>\beta,\leq\delta}}{OPM_{j,q}^{>\beta,\leq\delta}}$

and

$\qquad\qquad -\dfrac{OPM_{j,q}^{\leq\beta,\leq\delta} - OPM_{i,j,q}^{\leq\alpha,\leq\beta,\leq\delta}}{OPM_{j,q}^{\leq\beta,\leq\delta}} \quad i,\,j,\,q = 1,2,\dots,n.$

$\qquad\quad MJBOPMFV_{ijq>} = \dfrac{OPM_{j,q}^{>\beta,>\delta} - OPM_{i,j,q}^{\leq\alpha,>\beta,>\delta}}{OPM_{j,q}^{>\beta,>\delta}}$

$\qquad\qquad -\dfrac{OPM_{j,q}^{\leq\beta,>\delta} - OPM_{i,j,q}^{\leq\alpha,\leq\beta,>\delta}}{OPM_{j,q}^{\leq\beta,>\delta}} \quad i,\,j,\,q = 1,2,\dots,n.$

Thus, $MJBOPMFV$ of three components i, j and q can be defined as,

$$MJBOPMFV_{ijq} = MJBOPMFV_{ijq>} - MJBOPMFV_{ijq\leq},\ i,j,q = 1,2,\dots,n.$$

To define joint risk IMs, define the following indexes in terms of system risk.

$F_i^+(t)$: Value of risk metric F when component i has been in state below a specified level throughout the time interval $[0, t]$.

$F_i^-(t)$: Value of risk metric F when component i has been in state above a specified level throughout the time interval $[0, t]$.

$F_{i,j}^{++}(t)$: Value of risk metric F when components i and j have been in state below some specified levels throughout the time interval $[0, t]$.

$F_{i,j}^{+-}(t)$: Value of risk metric F when components, i has been in state below some specified level and j has been in state above some specified level, throughout the time interval $[0, t]$.

$F_{i,j}^{-+}(t)$: Value of risk metric F when components, i has been in state above some specified level and j has been in state below some specified level, throughout the time interval $[0, t]$.

$F_{i,j}^{--}(t)$: Value of risk metric F when both components i and j have been in state above some specified levels throughout the time interval $[0, t]$.

$F_{i,j,q}^{+++}(t)$: Value of risk metric F when the components i, j and q have been in state below some specified levels throughout the time interval $[0, t]$.

$F_{i,j,q}^{++-}(t)$: Value of risk metric F when components, i and j have been in state below some specified level and q has been in state above some specified level, throughout the time interval $[0, t]$.

$F_{i,j,q}^{+-+}(t)$: Value of risk metric F when components, j has been in state above some specified level and i and q have been in state below some specified level, throughout the time interval $[0, t]$.

$F_{i,j,q}^{-++}(t)$: Value of risk metric F, when components, i has been in state above some specified level and j and q have been in state below some specified level, throughout the time interval $[0, t]$.

$F_{i,j,q}^{--+}(t)$: Value of risk metric F when components, i and j have been in state above some specified level and q has been in state below some specified level, throughout the time interval $[0, t]$.

$F_{i,j}^{+--}(t)$: Value of risk metric F when components, j and q have been in state above some specified level and i has been in state below some specified level, throughout the time interval $[0, t]$.

$F_{i,j}^{-+-}(t)$: Value of risk metric F when components, i and q have been in state above some specified level and j has been in state below some specified level, throughout the time interval $[0, t]$.

$F_{i,j,q}^{---}(t)$: Value of risk metric F when components i, j and q have been in state above some specified levels throughout the time interval $[0, t]$.

Now define joint risk IMs with reference to three components i, j and q.

Birnbaum risk IM: $rB_i(t) = F_i^+(t) - F_i^-(t)$ measures the maximum deviation of risk when ith component shifts from higher level to lower level. Joint risk IMs for three components are as follows:

a. Multistate Joint Birnbaum-Risk Achievement Worth (*MJBrAW*)

$$Let \quad MJBrAW_{ijq+} = \frac{F_{ijq}^{+++}(t)}{F_{jq}^{++}(t)} - \frac{F_{ijq}^{++-}(t)}{F_{jq}^{-+}(t)} \quad and \quad MJBrAW_{ijq-} = \frac{F_{ijq}^{+--}(t)}{F_{jq}^{+-}(t)} - \frac{F_{ijq}^{+--}(t)}{F_{jq}^{--}(t)}$$

represent the *MJBrAW* of two components i and j when third component q is in below and above a prespecified threshold, respectively.

Thus, *MJBrAW* of three components i, j and q can be defined as,

$$MJBrAW_{ijq} = MJBrAW_{ijq+} - MJBrAW_{ijq-} \text{ where } i, \ j, \ q = 1,2,\ldots,n.$$

It is the change in *rAW* of component i when interaction performance of component j and q switches from high level to low level.

b. Multistate Joint Birnbaum-Risk Reduction Worth (*MJBrRW*)

$$Let \ MJBrRW_{ijq+} = \frac{F_{jq}^{++}(t)}{F_{ijq}^{++}(t)} - \frac{F_{jq}^{-+}(t)}{F^{--+}_{ijq}(t)} \quad and \quad MJBrRW_{ijq-} = \frac{F_{jq}^{+-}(t)}{F_{ijq}^{+-}(t)} - \frac{F_{jq}^{--}(t)}{F^{---}_{ijq}(t)}$$

represents the *MJBrRW* of two components i and j when third component q is in below and above a prespecified threshold, respectively.

Thus, *MJBrRW* of three components i, j and q can be defined as,

$$JBrRW_{ijq} = MJBrRW_{ijq+} - MJBrRW_{ijq-} \text{ where } i, \ j, \ q = 1,2,\ldots,n.$$

It is the change in *rRW* of component i when interaction performance of component j and q switches from high level to low level.

c. Multistate Joint Birnbaum Risk Fussell-Vesely measure (*MJBrFV*)

$$Let \ MJBrFV_{ijq+} = \frac{F_{jq}^{++}(t) - F_{ijq}^{-++}(t)}{F_{jq}^{++}(t)} - \frac{F_{jq}^{-+}(t) - F_{ijq}^{--+}(t)}{F_{jq}^{-+}(t)} \quad and$$

$$MJBrFV_{ijq-} = \frac{F_{jq}^{+-}(t) - F_{ijq}^{-+-}(t)}{F_{jq}^{+-}(t)} - \frac{F_{jq}^{-+}(t) - F_{ijq}^{--+}(t)}{F_{jq}^{-+}(t)} \quad represent \ the \ MJBrFV$$

of two components i and j when third component q is in below and above a prespecified threshold, respectively.

Thus, *MJBrFV* of three components i, j and q can be defined as,

$$MJBrFV_{ijq-} = MJBrFV_{ijq+} - MJBrFV_{ijq-} \text{ where } i, \ j, \ q = 1,2,\ldots,n.$$

It is the change in *rAW* of component i when interaction performance of components j and q switches from high level to low level.

9.4   EVALUATION PROCEDURE

A component's performance restriction approach for computation of the joint IMs and the evaluation procedure using UGF is given below.

9.4.1   STATE SPACE RESTRICTION APPROACH

Let OPM_{ik} be the OPM (reliability, expected system performance or risk) of the MSS when component i is in a fixed state k while the rest of components stay randomly in their corresponding states with steady-state performance distributions $\{x_{jl}, p_{jl}\}, 1 \le j \le n, 0 \le l \le M_j$.

The conditional probabilities of the component i being in state k with performance x_{ik} not greater than a pre-specified level threshold α $(k \le k_{i\alpha})$ and not less than a pre-specified level threshold $\alpha(k > k_{i\alpha})$ are

$$\Pr[X_i = x_{ik} \mid k \le k_{i\alpha}] = p^*_{1ik} = \frac{p_{ik}}{\sum_{r=0}^{k_{i\alpha}} p_{ir}} = \frac{p_{ik}}{p^{\le \alpha}_i} \text{ and } \Pr[X_i = x_{ik} \mid k > k_{i\alpha}]$$

$$= p^*_{2ik} = \frac{p_{ik}}{\sum_{r=k_{i\alpha}+1}^{M_i} p_{ir}} = \frac{p_{ik}}{p^{> \alpha}_i}$$

The joint probability distribution of two independent components i and j, for $X_i = x_{ik}, X_j = x_{jh}$, given four additional restrictions, (1) $k > k_{i\alpha}, h > h_{j\beta}$, (2) $k \le k_{i\alpha}, h > h_{j\beta}$, (3) $k > k_{i\alpha}, h \le h_{j\beta}$ and (4) $k \le k_{i\alpha}, h \le h_{j\beta}$, is

$$\Pr[X_i = x_{ik}, X_j = x_{jh} \mid k \le k_{i\alpha}, h \le h_{j\beta}] = p_1^{**}{}_{kh} = \frac{p_{ik} p_{jh}}{\sum_{r=0}^{k_{i\alpha}} p_{ir} \sum_{m=0}^{h_{j\beta}} p_{jm}},$$

$$\Pr[X_i = x_{ik}, X_j = x_{jh} \mid k \le k_{i\alpha}, h > h_{j\beta}] = p_2^{**}{}_{kh} = \frac{p_{ik} p_{jh}}{\sum_{r=0}^{k_{i\alpha}} p_{ir} \sum_{m=h_{j\beta}+1}^{M_j} p_{jm}},$$

$$\Pr[X_i = x_{ik}, X_j = x_{jh} \mid k > k_{i\alpha}, h \le h_{j\beta}] = p_3^{**}{}_{kh} = \frac{p_{ik} p_{jh}}{\sum_{r=k_{i\alpha}+1}^{M_i} p_{ir} \sum_{m=0}^{h_{j\beta}} p_{jm}},$$

$$\text{and } \Pr[X_i = x_{ik}, X_j = x_{jh} \mid k > k_{i\alpha}, h > h_{j\beta}] = p_4^{**}{}_{kh} = \frac{p_{ik} p_{jh}}{\sum_{r=k_{i\alpha}+1}^{M_i} p_{ir} \sum_{m=h_{j\beta}+1}^{M_j} p_{jm}}.$$

Similarly, joint probability distribution of three components i, j and q, for $X_i = x_{ik}$, $X_j = x_{jh}, X_q = x_{qs}$, given eight additional restrictions (1) $k \le k_{i\alpha}, h \le h_{j\beta}, s \le s_{q\delta}$, (2) $k \le k_{i\alpha}, h \le h_{j\beta}, s > s_{q\delta}$, (3) $k \le k_{i\alpha}, h > h_{j\beta}, s > s_{q\delta}$, (4) $k \le k_{i\alpha}, h > h_{j\beta}, s \le s_{q\delta}$, (5) $k > k_{i\alpha}, h \le h_{j\beta}, s \le s_{q\delta}$, (6) $k > k_{i\alpha}, h \le h_{j\beta}, s > s_{q\delta}$, (7) $k > k_{i\alpha}, h > h_{j\beta}, s > s_{q\delta}$, (8) $k > k_{i\alpha}, h > h_{j\beta}, s \le s_{q\delta}$ is

$$P\left[X_i = x_{ik}, X_j = x_{jh}, X_q = x_{qs} \mid k \le k_{i\alpha}, h \le h_{j\beta}, s \le s_{q\delta}\right]$$

$$= p_1^{***}{}_{khs} = \frac{p_{ik} p_{jh} p_{qs}}{\sum_{r=0}^{k_{i\alpha}} p_{ir} \sum_{m=0}^{h_{j\beta}} p_{jm} \sum_{l=0}^{s_{q\delta}} p_{ql}}$$

$$P\left[X_i = x_{ik}, X_j = x_{jh}, X_q = x_{qs} \mid k \le k_{i\alpha}, h \le h_{j\beta}, s > s_{q\delta}\right]$$

$$= p_2^{***}{}_{khs} = \frac{p_{ik} \, p_{jh} \, p_{qs}}{\sum_{r=0}^{k_{i\alpha}} p_{ir} \, \sum_{m=0}^{h_{j\beta}} p_{jm} \, \sum_{l=s_{q\delta}+1}^{M_l} p_{ql}}$$

$$P\left[X_i = x_{ik}, X_j = x_{jh}, X_q = x_{qs} \mid k \le k_{i\alpha}, h > h_{j\beta}, s > s_{q\delta}\right]$$

$$= p_3^{***}{}_{khs} = \frac{p_{ik} \, p_{jh} \, p_{qs}}{\sum_{r=0}^{k_{i\alpha}} p_{ir} \, \sum_{m=h_{j\beta}}^{M_m} p_{jm} \, \sum_{l=s_{q\delta}+1}^{M_l} p_{ql}}$$

$$P\left[X_i = x_{ik}, X_j = x_{jh}, X_q = x_{qs} \mid k \le k_{i\alpha}, h > h_{j\beta}, s \le s_{q\delta}\right]$$

$$= p_4^{***}{}_{khs} = \frac{p_{ik} \, p_{jh} \, p_{qs}}{\sum_{r=0}^{k_{i\alpha}} p_{ir} \, \sum_{m=h_{j\beta}}^{M_m} p_{jm} \, \sum_{l=0}^{s_{q\delta}} p_{ql}}$$

$$P\left[X_i = x_{ik}, X_j = x_{jh}, X_q = x_{qs} \mid k > k_{i\alpha}, h \le h_{j\beta}, s \le s_{q\delta}\right]$$

$$= p_5^{***}{}_{khs} = \frac{p_{ik} \, p_{jh} \, p_{qs}}{\sum_{r=k_{i\alpha}+1}^{M_r} p_{ir} \, \sum_{m=0}^{h_{j\beta}} p_{jm} \, \sum_{l=0}^{s_{q\delta}} p_{ql}}$$

$$P\left[X_i = x_{ik}, X_j = x_{jh}, X_q = x_{qs} \mid k > k_{i\alpha}, h \le h_{j\beta}, s > s_{q\delta}\right]$$

$$= p_6^{***}{}_{khs} = \frac{p_{ik} \, p_{jh} \, p_{qs}}{\sum_{r=k_{i\alpha}+1}^{M_r} p_{ir} \, \sum_{m=0}^{h_{j\beta}} p_{jm} \, \sum_{l=s_{q\delta}+1}^{M_l} p_{ql}}$$

$$P\left[X_i = x_{ik}, X_j = x_{jh}, X_q = x_{qs} \mid k > k_{i\alpha}, h > h_{j\beta}, s > s_{q\delta}\right]$$

$$= p_7^{***}{}_{khs} = \frac{p_{ik} \, p_{jh} \, p_{qs}}{\sum_{r=k_{i\alpha}+1}^{M_r} p_{ir} \, \sum_{m=h_{j\beta}}^{M_m} p_{jm} \, \sum_{l=s_{q\delta}+1}^{M_l} p_{ql}}$$

$$P\left[X_i = x_{ik}, X_j = x_{jh}, X_q = x_{qs} \mid k > k_{i\alpha}, h > h_{j\beta}, s \le s_{q\delta}\right]$$

$$= p_8^{***}{}_{khs} = \frac{p_{ik} \, p_{jh} \, p_{qs}}{\sum_{r=k_{i\alpha}+1}^{M_r} p_{ir} \, \sum_{m=h_{j\beta}}^{M_m} p_{jm} \, \sum_{l=0}^{s_{q\delta}} p_{ql}}$$

Using the above conditional probability distributions, OPMs are

$$OPM^{\le\alpha}{}_i = \sum_{k=0}^{k_{i\alpha}} \frac{p_{ik}}{p^{\le\alpha}{}_i} OPM_{ik}, OPM^{>\alpha}{}_i = \sum_{k=k_{i\alpha}+1}^{M_i} \frac{p_{ik}}{p^{>\alpha}{}_i} OPM_{ik},$$

$$OPM_{ij}^{\le\alpha,\le\beta} = \sum_{k=0}^{k_{i\alpha}} \sum_{h=0}^{h_{j\beta}} p_1^{**}{}_{kh} OPM_{ik.jh}, OPM_{ij}^{\le\alpha,>\beta} = \sum_{k=0}^{k_{i\alpha}} \sum_{h=h_{j\beta}+1}^{M_j} p_2^{**}{}_{kh} OPM_{ik.jh},$$

$$OPM_{ij}^{>\alpha,\le\beta} = \sum_{k=k_{i\alpha}+1}^{M_i} \sum_{h=0}^{h_{j\beta}} p_3^{**}{}_{kh} OPM_{ik.jh}, OPM_{ij}^{>\alpha,>\beta} = \sum_{k=k_{i\alpha}+1}^{M_i} \sum_{h=k_{j\beta}+1}^{M_j} p_4^{**}{}_{kh} OPM_{ik.jh}$$

$$OPM_{i,j,q}^{>\alpha,>\beta,>\delta} = \sum_{k=k_{i\alpha}+1}^{M_i} \sum_{h=h_{j\beta}+1}^{M_j} \sum_{s=s_{q\delta}+1}^{M_q} OPM_{ik,jh,qs} p_1^{***}{}_{khs}$$

$$OPM_{i,j,q}^{>\alpha,\leq\beta,>\delta} = \sum_{k_{i\alpha}+1}^{M_i} \sum_{h=0}^{h_{j\beta}} \sum_{s_{q\delta}+1}^{M_q} OPM_{ik,jh,qs} \overset{***}{p_2}_{khs}$$

$$OPM_{i,j,q}^{>\alpha,>\beta,\leq\delta} = \sum_{k=k_{i\alpha}+1}^{M_i} \sum_{h=h_{j\beta}+1}^{M_j} \sum_{s=0}^{s_{q\delta}} OPM_{ik,jh,qs} \overset{***}{p_3}_{khs}$$

$$OPM_{i,j,q}^{>\alpha,>\beta,\leq\delta} = \sum_{k=k_{i\alpha}+1}^{M_i} \sum_{h=0}^{h_{j\beta}} \sum_{s=0}^{s_{q\delta}} OPM_{ik,jh,qs} \overset{***}{p_4}_{khs}$$

$$OPM_{i,j,q}^{\leq\alpha,>\beta,>\delta} = \sum_{k=0}^{k=k_{i\alpha}} \sum_{h=h_{j\beta}+1}^{M_j} \sum_{s=s_{q\delta}+1}^{M_q} OPM_{ik,jh,qs} \overset{***}{p_5}_{khs}$$

$$OPM_{i,j,q}^{\leq\alpha,\leq\beta,>\delta} = \sum_{k=0}^{k=k_{i\alpha}} \sum_{h=0}^{h_{j\beta}} \sum_{s_{q\delta}+1}^{M_q} OPM_{ik,jh,qs} \overset{***}{p_6}_{khs}$$

$$OPM_{i,j,q}^{\leq\alpha,>\beta,\leq\delta} = \sum_{k=0}^{k=k_{i\alpha}} \sum_{h=h_{j\beta}+1}^{M_j} \sum_{s=0}^{s_{q\delta}} OPM_{ik,jh,qs} \overset{***}{p_7}_{khs}$$

$$OPM_{i,j,q}^{\leq\alpha,>\beta,\leq\delta} = \sum_{k=0}^{k=k_{i\alpha}} \sum_{h=0}^{h_{j\beta}} \sum_{s=0}^{s_{q\delta}} OPM_{ik,jh,qs} \overset{***}{p_8}_{khs}$$

where OPM_{ik}, $OPM_{ik,jh}$ and $OPM_{ik,jh,qs}$ are the system steady-state OPMs when component i is in state k, component i is in state k and component j is in state h, and component i is in state k, component j is in state h and component q is in state s, while the rest of components stay randomly in their corresponding states with steady-state performance distributions $\{x_{jl}, p_{jl}\}$, $1 \leq j \leq n$, $0 \leq l \leq M_j$.

9.4.2 UNIVERSAL GENERATING FUNCTION TECHNIQUE

Ushakov (1986) introduced UGF. It is a very effective tool for the reliability evaluation of different types of MSSs. To find state space restriction, use the u-functions of components i, j and q as follows:

$$u_i^{\leq\alpha}(z) = \sum_{r=0}^{k_{i\alpha}} \left(p_{ir}/p_i^{\leq\alpha} \right) z^{x_{ir}} \text{ for } OPM_i^{\leq\alpha}, \; u_i^{>\alpha}(z) = \sum_{r=k_{i\alpha}+1}^{M_i} \left(p_{ir}/p_i^{>\alpha} \right) z^{x_{ir}} \text{ for } OPM_i^{>\alpha},$$

$$u_{i,j}^{\leq\alpha,\leq\beta}(z) = \sum_{r=0}^{k_{kz}} \left(p_{ir}/p_i^{\leq\alpha} \right) z^{x_{ir}} * \sum_{m=0}^{k_{j\beta}} \left(p_{jm}/p_j^{\leq\beta} \right) z^{x_{jm}}$$

$$= \sum_{r=0}^{k_{i\alpha}} \sum_{m=0}^{k_{j\beta}} \left(p_{ir} p_{jm}/p_i^{\leq\alpha} p_j^{\leq\beta} \right) z^{\omega(x_{ir}, x_{jm})} \text{ for } OPM_{i,j}^{\leq\alpha,\leq\beta}$$

$$u_{i,j}^{>\alpha,\leq\beta}(z) = \sum_{r=k_{i\alpha}}^{M_i} \left(p_{ik}/p_i^{>\alpha} \right) z^{x_{ik}} * \sum_{m=0}^{k_{j\beta}} \left(p_{jm}/p_j^{\leq\beta} \right) z^{x_{jm}}$$

$$= \sum_{r=k_{i\alpha}}^{M_i} \sum_{m=0}^{k_{j\beta}} \left(p_{ir} p_{jm}/p_i^{>\alpha} p_j^{\leq\beta} \right) z^{\omega(x_{ir}, x_{jm})} \text{ for } OPM_{i,j}^{>\alpha,\leq\beta}$$

$$u_{i,j}^{\leq\alpha,>\beta}(z) = \sum_{r=0}^{k_{j\alpha}} \left(p_{ir}/p_i^{\leq\alpha} \right) z^{x_{ir}} * \sum_{m=k_{j\beta}+1}^{M_j} \left(p_{jm}/p_j^{>\beta} \right) z^{x_{jm}}$$

$$= \sum_{r=0}^{k_{i\alpha}} \sum_{m=k_{j\beta}+1}^{M_j} \left(p_{ir}p_{jm}/p_i^{\leq\alpha}p_j^{>\beta} \right) z^{\omega\left(x_{ir},x_{jm}\right)} \text{ for } OPM_{i,j}^{\leq\alpha,>\beta} \text{ and}$$

$$u_{i,j}^{>\alpha,>\beta}(z) = \sum_{r=k_{i\alpha}+1}^{M_i} \left(\frac{p_{ir}}{p_i^{>\alpha}} \right) z^{x_{ir}} * \sum_{m=k_{j\beta}+1}^{m_j} \left(\frac{p_{jm}}{p_j^{>\beta}} \right) z^{x_{jm}}$$

$$= \sum_{r=k_{i\alpha}+1}^{M_i} \sum_{m=k_{j\beta}+1}^{M_j} \left(\frac{p_{ir}p_{jm}}{p_i^{>\alpha}p_j^{>\beta}} \right) z^{\omega\left(x_{ir},x_{jm}\right)} \text{ for } OPM_{i,j}^{>\alpha,>\beta},$$

$$u_{i,j,q}^{>\alpha,>\beta,>\delta}(z) = \sum_{k=k_{i\alpha}+1}^{M_i} \left(\frac{p_{ik}}{p_i^{>\alpha}} \right) z^{x_{ik}} * \sum_{h=h_{j\beta}+1}^{M_j} \left(\frac{p_{jh}}{p_j^{>\beta}} \right) z^{x_{jh}} * \sum_{s=s_{q\delta}+1}^{M_q} \left(\frac{p_{qs}}{p_q^{>\delta}} \right) z^{x_{qs}}$$

$$u_{i,j,q}^{>\alpha,>\beta,>\delta}(z) = \sum_{k=k_{i\alpha}+1}^{M_i} \sum_{h=h_{j\beta}+1}^{M_j} \sum_{s=s_{q\delta}+1}^{M_q} p_1^{***}{}_{khs} z^{\omega\left(x_{ik},x_{jh},x_{qs}\right)}$$

$$u_{i,j,q}^{>\alpha,>\beta,>\delta}(z) = u_{i,j,q}^{>\alpha,>\beta,\leq\delta}(z) = \sum_{k=k_{i\alpha}+1}^{M_i} \left(\frac{p_{ik}}{p_i^{>\alpha}} \right) z^{x_{ik}} * \sum_{h=h_{j\beta}+1}^{M_j} \left(\frac{p_{jh}}{p_j^{>\beta}} \right) z^{x_{jh}}$$

$$* \sum_{s=0}^{s_{q\delta}} \left(\frac{p_{qs}}{p_q^{>\delta}} \right) z^{x_{qs}} \sum_{k=k_{i\alpha}+1}^{M_i} \sum_{h=h_{j\beta}+1}^{M_j} \sum_{s=s_{q\delta}+1}^{M_q} p_1^{***}{}_{khs} z^{\omega\left(x_{ik},x_{jh},x_{qs}\right)}$$

$$u_{i,j,q}^{>\alpha,>\beta,\leq\delta}(z) = \sum_{k=k_{i\alpha}+1}^{M_i} \sum_{h=h_{j\beta}+1}^{M_j} \sum_{s=0}^{s_{q\delta}} p_1^{***}{}_{khs} z^{\omega\left(x_{ik},x_{jh},x_{qs}\right)}$$

$$u_{i,j,q}^{>\alpha,\leq\beta,>\delta}(z) = \sum_{k=k_{i\alpha}+1}^{M_i} \left(\frac{p_{ik}}{p_i^{>\alpha}} \right) z^{x_{ik}} * \sum_{h=0}^{h=h_{j\beta}} \left(\frac{p_{jh}}{p_j^{>\beta}} \right) z^{x_{jh}} * \sum_{s=s_{q\delta}+1}^{M_q} \left(\frac{p_{qs}}{p_q^{>\delta}} \right) z^{x_{qs}}$$

$$u_{i,j,q}^{>\alpha,\leq\beta,>\delta}(z) = \sum_{k=k_{i\alpha}+1}^{M_i} \sum_{h=0}^{h=h_{j\beta}} \sum_{s=s_{q\delta}+1}^{M_q} p_1^{***}{}_{khs} z^{\omega\left(x_{ik},x_{jh},x_{qs}\right)}$$

$$u_{i,j,q}^{>\alpha,\leq\beta,\leq\delta}(z) = \sum_{k=k_{i\alpha}+1}^{M_i} \left(\frac{p_{ik}}{p_i^{>\alpha}} \right) z^{x_{ik}} * \sum_{h=0}^{h=h_{j\beta}} \left(\frac{p_{jh}}{p_j^{>\beta}} \right) z^{x_{jh}} * \sum_{s=0}^{s=s_{q\delta}} \left(\frac{p_{qs}}{p_q^{>\delta}} \right) z^{x_{qs}}$$

$$u_{i,j,q}^{>\alpha,\leq\beta,\leq\delta}(z) = \sum_{k=k_{i\alpha}+1}^{M_i} \sum_{h=0}^{h=h_{j\beta}} \sum_{s=0}^{s=s_{q\delta}} p_1^{***}{}_{khs} z^{\omega\left(x_{ik},x_{jh},x_{qs}\right)}$$

Similarly

$$u_{i,j,q}^{\le\alpha,>\beta,>\delta}(z) = \sum_{k=0}^{k=k_{i\alpha}}\left(\frac{p_{ik}}{p_i^\alpha}\right)z^{x_{ik}} * \sum_{h=h_{j\beta}+1}^{M_j}\left(\frac{p_{jh}}{p_j^\beta}\right)z^{x_{jh}} * \sum_{s=s_{q\delta}+1}^{M_q}\left(\frac{p_{qs}}{p_q^\delta}\right)z^{x_{qs}}$$

$$u_{i,j,q}^{\le\alpha,>\beta,>\delta}(z) = \sum_{k=0}^{k=k_{i\alpha}}\sum_{h=h_{j\beta}+1}^{M_j}\sum_{s=s_{q\delta}+1}^{M_q} p_1^{***}{}_{khs}z^{\omega(x_{ik},x_{jh},x_{qs})}$$

$$u_{i,j,q}^{\le\alpha,>\beta,\le\delta}(z) = \sum_{k=0}^{k=k_{i\alpha}}\left(\frac{p_{ik}}{p_i^\alpha}\right)z^{x_{ik}} * \sum_{h=h_{j\beta}+1}^{M_j}\left(\frac{p_{jh}}{p_j^\beta}\right)z^{x_{jh}} * \sum_{s=0}^{s_{q\delta}}\left(\frac{p_{qs}}{p_q^\delta}\right)z^{x_{qs}}$$

$$u_{i,j,q}^{\le\alpha,\le\beta,>\delta}(z) = \sum_{k=0}^{k=k_{i\alpha}}\left(\frac{p_{ik}}{p_i^\alpha}\right)z^{x_{ik}} * \sum_{h=0}^{h=h_{j\beta}}\left(\frac{p_{jh}}{p_j^\beta}\right)z^{x_{jh}} * \sum_{s=s_{q\delta}+1}^{M_q}\left(\frac{p_{qs}}{p_q^\delta}\right)z^{x_{qs}}$$

$$u_{i,j,q}^{\le\alpha,\le\beta,>\delta}(z) = \sum_{k=0}^{k=k_{i\alpha}}\sum_{h=0}^{h=h_{j\beta}}\sum_{s=s_{q\delta}+1}^{M_q} p_1^{***}{}_{khs}z^{\omega(x_{ik},x_{jh},x_{qs})}$$

$$u_{i,j,q}^{\le\alpha,\le\beta,\le\delta}(z) = \sum_{k=0}^{k=k_{i\alpha}}\left(\frac{p_{ik}}{p_i^\alpha}\right)z^{x_{ik}} * \sum_{h=0}^{h=h_{j\beta}}\left(\frac{p_{jh}}{p_j^\beta}\right)z^{x_{jh}} * \sum_{s=0}^{s=s_{q\delta}}\left(\frac{p_{qs}}{p_q^\delta}\right)z^{x_{qs}}$$

$$u_{i,j,q}^{>\alpha,\le\beta,\le\delta}(z) = \sum_{k=k_{i\alpha}+1}^{M_i}\sum_{h=0}^{h=h_{j\beta}}\sum_{s=0}^{s=s_{q\delta}} p_1^{***}{}_{khs}z^{\omega(x_{ik},x_{jh},x_{qs})}$$

then apply the UGF algorithm. The coefficients of above UGFs are used for the evaluation of joint IMs using OPMs.

9.5 ILLUSTRATIVE EXAMPLE

Consider a system made up of $n = 4$ multistate components in series logic. Each component has $M_j = 4$ nonzero states, 0, 1, 2, 3 and 4, with corresponding values of performance $x_{j0}=0$, $x_{j1}=25$, $x_{j2}=50$, $x_{j3}=75$, $x_{j4}=100$, $j=1, 2, 3, 4$ (see Figure 9.1).

Let p_{jk} be the probability of component j of being in state k and are given in Table 9.1. By considering states 0, 1 and 2 (unreliable states) for $\le\alpha$ or $\le\beta$ or $\le\delta$ and states 3 and 4 (reliable states) for $>\alpha$ or $>\beta$ or $>\delta$, the Multistate Expected Output Performance Measure (MSEOPM) with restriction to reliable or unreliable states are obtained.

Multistate joint IMs are computed and given in Table 9.2 and plotted in Figures 9.2 and 9.3. A numerical comparison can be made using the sign and size of the value

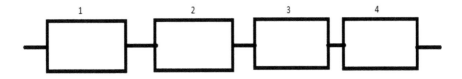

FIGURE 9.1   Series system.

TABLE 9.1
Probability Distributions of Components 1, 2, 3 and 4

	Component Number							
Probability Distribution	**1**		**2**		**3**		**4**	
$P(Xi0=0)$	p10	=0.5	p20	=0.45	p30	=0.4	p40	=0.45
$P(Xi1=25)$	p11	=0	p21	=0	p31	=0.1	p41	=0
$P(Xi2=50)$	p12	=0	p22	=0.1	p32	=0	p42	=0.1
$P(Xi3=75)$	p13	=0	p23	=0	p33	=0.1	p43	=0
$P(Xi4=100)$	p14	=0.5	p24	=0.45	p34	=0.4	p44	=0.45

TABLE 9.2
Multistate Joint-Performance Importance Measures

Joint importance measures for components 1, 2, 3

$MJBOPMAW_s=-6.80154$	$MJBOPMAW_s=21.4555$	$MJBOPMAW = 14.65399$
$MJBOPMRW_s=0.058304$	$MJBOPMRW_s=-0.09922$	$MJBOPMRW = 0.157525$
$MJBOPMFV_s=26.14749$	$MJBOPMRW_s=-2.21813$	$MJBOPMFV = 28.36562$

Joint importance measures for components 2, 3, 4

$MJBOPMAW_s=4.947823$	$MJBOPMAW_s=-3.02808$	$MJBOPMAW = 7.975904$
$MJBOPMRW_s=-0.60779$	$MJBEOPMRW_s=-4.34019$	$MJBOPMRW = 3.732395$
$MJBOPMFV_s=0.139011$	$MJBOPMRW_s=-4.22496$	$MJBOPMFV = 4.363972$

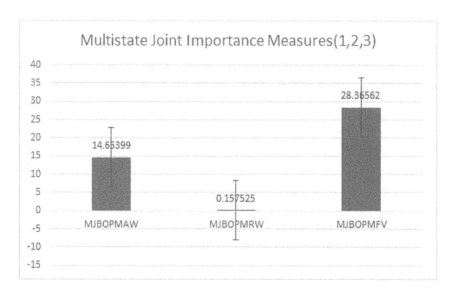

FIGURE 9.2   Multistate joint importance measures of components 1, 2 and 3.

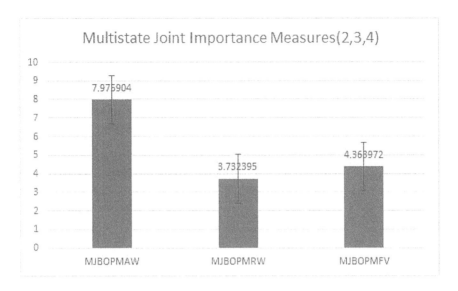

FIGURE 9.3 Multistate joint importance measures of components 2, 3 and 4.

of relevant measure with regard to their impact on expected system output performance. Since the *MJBOPMFV* = 28.36562 and *MJBOPMAW* = 14.65399 are much higher than *MJBOPMRW* = 0.157525, expected output performance achievement is higher due to the presence of high level of interaction effect of three components 1, 2 and 3. But expected output performance reduction due to low-level interaction effect of three components 1, 2 and 3 is very small.

Since *MJBOPMAW* = 7.975904 is higher than *MJBOPMRW* = 3.732395 and *MJBOPMFV* = 4.363972, expected output performance achievement is slightly better due to the presence of high level of interaction effect of three components 2, 3 and 4 while expected output performance reduction worth and FV measure due to low-level interaction effect of three components 2, 3 and 4 is small.

Highest value for *MJBEOPMAW* and *MJBEOPMFV* are attained for the group of components (1, 2 and 3) while highest value of *MJBEOPMRW* is attained to the group (2, 3 and 4). Highest values of *MJBOPMAW* and *MJBOPMFV* are due to highest change in IMs, *BOPMAW* and *BOPMFV*, respectively, of component 1 while the interaction performance of component 2 and 3 changes from higher level to lower level. Highest value of *MJBEOPMAW* is due to highest change in IM *BOPMAW* of component 2 while the interaction performance of component *3* and 4 changes from higher level to lower level.

This information can be used to provide more safety and redundancy operations for a group of components compared to other group. Highest values in various IMs indicate the need for more reliability and safety operations. A researcher needs to understand the variation to adopt reliability improvement activities and can give more care to group of components.

9.6   RESULT DISCUSSION

The present study proposed a measure, in binary and multistate context, that quantifies the maximum possible deviation in *RAW*, maximum change in *RRW* and maximum change in reliability *FV* measures due to a particular component i when interaction effect of component j and q switches from high level to low level, while both components j and q move from high level to low level simultaneously or individually, $i \neq j \neq q$, i, j, $q \in \{1, 2, ..., n\}$, n is the number of components in the system.

MJBRAW measures the joint effect of reliability achievement worth of component i and the interaction effect of components j and q, measured in Birnbaum sense, with reference to the performance measure reliability. Similarly, *MJBRRW* measures the joint effect of reliability reduction worth of component i and the interaction effect of components j and q, measured in Birnbaum sense. *MJBRFV* measures the joint effect of reliability *FV* of component i and the interaction of components j and q, measured in Birnbaum sense. When the performance measure changes, the proposed measures need to redefine. *MJBOPMAW*, *MJBOPMRW* and *MJBOPMFV* measures are also proposed for various OPMs, expected OPM, reliability, availability etc. To apply in risk-informed applications, *MJBrAW*, *MJBrRW* and *MJBrFV* measures are proposed.

The proposed measures can be used to prioritize reliability improvement activities on engineering systems.

9.7   CONCLUSION

This chapter introduced joint IMs for MSSs with reference to the OPMs' reliability, expected system performance and risk. The joint IMs for binary and MSSs, *JBRAW*, *JBRRW* and *JBRFV* for three components are introduced and generalized to various OPMs like expected OPM and availability. *MJBrAW*, *MJBrRW* and *MJBrFV* IMs are also introduced for three components. The new joint IMs are useful for prioritizing reliability improvement activities. The UGF method is used to evaluate the joint IMs, in which the system performance is measured in terms of productivity or capacity. Joint IM values will be useful for reliability design and maintenance activities.

ACKNOWLEDGMENT

The research is supported by "Santhome Research Grant (Seed Money)".

REFERENCES

R. E. Barlow and F. Proschan (1975), *Statistical Theory of Reliability and Life Testing*, New York, NY: Holt, Rinehart & Winston.

R. E. Barlow and A. Wu (1978), Coherent system with multistate components, *Mathematics and Operations Research*, Vol. 3, 275–281.

E. Borgonovo and G. E. Apostolakis (2001), A new importance measure for risk-informed decision making, *Reliability Engineering and System Safety*, Vol. 72, 193–212.

M. Van der Borst and H. Shoonakker (2001), An overview of PSA importance measures, *Reliability Engineering and System Safety*, Vol. 72, 241–245.

V. C. Bueno (1989), On the importance of components for multistate monotone systems, *Statistics and Probability Letters*, Vol. 7, 51–59.

V. M. Chacko and M. Manoharan (2008), Joint importance measures for the multistate system, *Advances in Performance and Safety of Complex systems*, A. K. Verma, P. K. Kapur, S. G. Ghadge, Eds., New Delhi: Macmillan, 308–314.

V. M. Chacko and M. Manoharan (2011), Joint importance measures for multistate reliability system, *Opsearch*, Vol. 48(3), 257–278.

V M Chacko (2020), New joint importance measures for multistate systems, *International Journal of Statistics and Reliability Engineering*, Vol. 7(1), 140–148.

V. M. Chacko (2021), On Birnbaum type joint importance measures for multistate reliability systems, *Communications in Statistics - Theory and Methods*, (Published Online) doi. 10.1080/03610926.2021.1961000

M. C. Cheok, G. W. Parry, and R. R. Sherry (1998), Use of importance measures in risk informed applications, *Reliability Engineering and System Safety*, Vol. 60, 213–226.

I. Ushakov (1986), A universal generating function, *Soviet Journal of Computer and Systems Sciences.*, Vol. 24, 37–39.

D. Vasseur, and M. Llory (1999), International survey on PSA figures of merit, *Reliability Engineering and System Safety*, Vol. 66, 261–274.

S. Wu (2005), Joint importance measures of multistate systems, *Computers and Industrial Engineering*, Vol. 49, 63–75.

S. Wu and L. Chan (2003), Performance utility – analysis of multistate systems, *IEEE Transactions on Reliability*, Vol. 52, 14–20.

E. Zio, M. Marella, and L. Podofillini (2007), Importance measures-based prioritization for improving the performance of multi-state systems: Application to the railway industry. *Reliability Engineering and System Safety*, Vol. 92, 1303–1314.

10 Reliability Approach for Enhancing System's Performance

Fatemeh Afsharnia
Department of Agricultural Machinery and Mechanization Engineering, Agricultural Sciences and Natural Resources University of Khuzestan, Ahvaz, Iran

Notation

Main Acronyms	
TEEP	Total Effective Equipment Performance
FT	Fault tree
BNs	Bayesian networks
AND-gates	The AND gate is a basic digital logic gate which implements logical conjunction ($\wedge$) from mathematical logic
OR-gates	The OR gate is a digital logic gate which implements logical disjunction ($\vee$) from mathematical logic
DAGs	Directed acyclic graph
Main Symbols	
R_{system}	The reliability of system
S_r	The realized production speed
S_p	The planned production speed

CONTENTS

DOI: 10.1201/9781003156291-10

10.1 INTRODUCTION

Technology advancements, the spread of modern industry all over the world, increased investment in the industries, and the restricted availability of raw and processed materials are just some of the factors that have made performance, efficiency, and productivity one of the most important business issues for industry owners and managers. For any mechanism, efficiency means that the highest work efficiency is obtained from the mechanism by spending the shortest time interval, the minimum work, and the least equipment. The importance and merit of this goal will be significantly enhanced when a comprehensive analysis of industrial competition in global and domestic markets is presented.

The basis of these competitions is on huge changes in investments, an increase of production capacities, the increasing use of automatic systems, and the existence of production conditions that have led institutions to maintain product performance and withstand effects of the economic fluctuations. So, they always seek the correct and efficient techniques to improve their management skills and create the highest efficiency and profit for their institution and industry. Achieving this important goal is obtained for industry owners when they deal with production issues by a logical and accurate view as well as proper planning.

One of the main manufacturing challenges is equipment and machine maintenance operation and production facilities. The most basic solution is to use the planning and repair system, especially the preventive repair and maintenance strategy. In this system, the production or services equipment and machines are inspected after a specified period of the operation time. The number and time between inspections and repairs depending on the structure and the operating conditions of the equipment prevent wear and tear in unusual conditions and make the necessary decisions, and predictions for purchasing or manufacturing of spare parts required.

Maintenance is a crucial part for increasing the efficiency and performance of production and processing equipment in a motorized and mechanized fleet. A proper maintenance strategy should be able to minimize timelines costs by predicting reliability and planning to maintain and upgrade it.

Today, the machinery costs are a significant portion of the total fixed costs [1], for this reason, reliability, availability, and reduction of downtime and repair of equipment related to industries are doubly important, and the operators of this equipment have always tried to reduce the repair operations caused by the breakdown of their machines. Maintenance is used as a principle of global scale in manufacturing enterprises. In the agricultural sector, machine maintenance costs make up a significant portion of their total operating costs and depending on the type of machinery. These costs can be between 15 and 60% of the production cost [2].

Therefore, balancing maintenance and repair costs with machine reliability should be achievable in an optimal manner. A machine is susceptible to damage due to its unreliability during its entire life. Given that in developed countries the cost of repairs is 8% and in developing countries, which are mostly importers of technology and machinery from developed countries, approximately 53% of the total cost of equipment is spent on repairing its breakdowns [3]. It is necessary to carry out the maintenance and repairs studies using advanced techniques and methods to maximize the performance of the capital that is spent on the purchase of machines. Using lifetime optimization methods, the durability and resilience of equipment under preventive maintenance can be increased by 30–40% when compared to equipment under poor maintenance [4].

The high costs of repairs, replacement of parts, and maintenance of these machines have led the maintenance engineers to use modern management methods optimally and effectively. In most industries, the equipment is repairable, and the reliability of the system will decrease and the failure rate will increase by more using equipment as well as increasing their age. Therefore, it seems necessary to pay attention to the possibility of failure, repair time and its causes, and the use of accurate forecasting tools. Recently, the production process has changed due to the timely production approach in production management and operations, and the importance of continuous production has increased. Therefore, manufacturing companies must seek to increase efficiency and optimize their production operations planning to maintain and survive.

10.2 PERFORMANCE

Performance is doing everything in a certain way by setting a specific goal for it. Also, performance is defined as the process of explaining the quality, effectiveness, and efficiency of past operations. According to this definition, performance is divided into two parts: (1) **Efficiency** that describes how an organization uses resources in the services or production; (2) **Effectiveness** that describes the achievement degree of organizational goals. A true efficiency indicator of your manufacturing operation is the Total Effective Equipment Performance (TEEP). TEEP measures how effectively your manufacturing operation is utilizing its resources. TEEP is calculated as follows:

$$TEEP = Availability \times Performance \times Quality \times Utilization \qquad (10.1)$$

$$Performance = \frac{S_r}{S_p} \qquad (10.2)$$

where S_r is the realized production speed and S_p is the planned production speed.

10.3 EFFECT OF RELIABILITY ON PERFORMANCE

As it turns out, performance is derived from the ratio of realized production speed to the planned production speed. If the equipment of the production line breaks down, the realized production speed is reduced, and subsequently, the performance is declined. So, the equipment failures must be minimized. Reliability is the possibility of healthy and trouble-free operation of the system for a specified period and under specified and predetermined conditions. Reliability is used to describe operations and tasks and to measure performance in general.

10.4 MEASURING METHODS OF EQUIPMENT RELIABILITY

In this section, we will look at how to obtain the reliability of machines, components, and operations. The best way of the data record is based on these units. Fortunately, computers and spreadsheet software make this much easier today. Foremen, mechanics, or service managers can record the downtimes of repairs. For agricultural machinery, this is certainly possible. One of the largest sugarcane growers in the world tracked every farm machine and recorded its lifetime data via a computer. A farmer can record working time and overhaul data on spreadsheet software.

In the processing industry, maintenance data is often available, and data can be estimated by the recovery of past data related to items such as mixers, conveyors, packaging machines, and chemical processes. Where there is no lifetime data, a good manager can collect sampling data using good statistical techniques. Data can sometimes be collected from other parts industries or engineering companies. This includes some reliability data in processing journals, and when all the components fail, the farm systems manager can talk to machine and process operators to find breakdowns and downtime for repairs. Obtaining accurate data on unit reliability requires planning. However, trying to obtain information will result in high system performance.

10.4.1 ESTIMATING OF THE OPERATOR RELIABILITY

Many situational factors affect human function. The operator reliability can vary from zero to 0.9999. Therefore, the development of generalized communications will be quite difficult. Considering the exchanges that can be used are ultimately qualitative. For example, the human role is often affected by costs, risks, technology status, and other factors in an agricultural system. The reliability of each operator rarely exceeds 0.78 in the long run, because there are a lot of time-reducing values. Ignoring time losses on weekends and hours over 40 hours per week, time losses include:

- ✓ Time of illness
- ✓ Delay time
- ✓ Holiday time
- ✓ Vacation
- ✓ Time of breakdowns
- ✓ Other cases

People are also prone to making mistakes over and over again. The amount allocated to a person is highly subject to change depending on how people are entered into the system. These values can be greatly increased by proper planning and support. Let us take a closer look at the true value of a human in an agricultural system. Consider a female worker who operates a packing machine on a production line of garden mulch packages. If she works 7 hours a day, 6 days a week in a working year, she must be at work 2184 hours a year. But she will usually not be able to be present at all hours. Consider the following time losses in Table 10.1.

TABLE 10.1
Time Losses for a Female Worker

Activity	Time Losses
Two weeks' vacation	84
Holidays (15 days)	105
Leave (6 days)	42
Work breaks (30 minutes/day)	156
Delay (1 hour/week)	50
Illness (7 days)	42
All-time losses	479 hours

$$All\ time\ losses\ (\%) = \frac{479}{2184} \times 100 = 21.93\%$$

$$Reliability \cong 78.067\%$$

The first reaction is not to accept it, but this is only for one person. It will be easy for a worker to develop a broader scenario. So, how does an industry manager handle line workers who may include 20 or more workers?

First, it must be assumed that the manager can support this worker by employee backup. This manager may call someone else during his or her turn-off, or some workers might be present at his location. Its calculations will be as follows:

$$R_{Packing} = 0.7806 + 0.7806(1 - 0.7806) = 0.9518$$

It is observed that by having a backup, the reliability of each operator increases up to 95.18%. In some cases, people may be replaced by cars with more reliability, or some people may be replaced by cars, which seems to cost less.

If the manager supports this worker by two employee backups, reliability will increase dramatically. It will calculate as follows:

$$R_{packing} = 1 - (1 - 0.7806)(1 - 0.7806) = 0.9894$$

Increasing the employee backups would indeed increase the reliability, but it imposes a relatively higher cost on the system, so an optimal value should be estimated depending on the system type.

10.4.2 RELIABILITY OF COMPOSITE SYSTEMS

A system usually consists of some components or elements, or some smaller systems or subsystems. The system reliability will affect by the structure of system components to each other and the dependencies between these components. If a manufacturing plant is considered as a system, the machines and other facilities

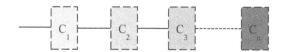

FIGURE 10.1 The components of a series system.

inside the workshop are considered as a set of components or subsystems that make up the workshop. In a machine-like system, the internal parts and components of the machine will be the subsystems of the machine [5].

10.4.2.1 Series Systems

Consider a workshop with a continuous production system. In this workshop, different machines, each with different tasks, are arranged in a series according to the needs of the production process. Here, the breakdowns of each machine will cause the stoppage of the workshop and subsequently the production process. The system components are arranged side by side or in series in Figure 10.1.

Most agricultural systems are arranged in series. The successful operation of a series system depends on the successful operation of every component in the system, including operator, machine, and process. There are often two conditions: (1) the failure of each component results in the failure of a composite system, and (2) component failures are independent of each other. In series systems, the probability of a system's reliable work is obtained from the reliability of the individual components. For example, if there are three components in a system, each with a reliability of 0.85, the reliability of the system results from the product of the reliability of all three components, which is equal to 0.614. The following equation is used to calculate the reliability of series combinations:

$$R_{system} = R_1 \times R_2 \times R_3 \times \cdots \times R_N \qquad (10.3)$$

where $R_1, R_2, \ldots, R_N$ are the reliability of the running components which are expressed as the probability of success in work based on history or estimates. The main weakness of most series systems is their inherent sequence, which leads to a decrease in the overall reliability of the system. So, for the series configuration, the overall reliability of the system is smaller than the reliability of each component.

Example 10.1
Consider a simple wheat harvesting system with a tractor (0.8) to pull a cart (0.95) that the combine (0.92) is discharged into a storage tank (Figure 10.2).

$$R_{system} = R_{cart} \times R_{tractor} \times R_{harvester} = 0.95 \times 0.8 \times 0.92 = 0.6992$$

In this case, the system reliability is calculated to be 0.6992. If one unit fails, the whole system stops.

As more units are added to an agricultural system, if it is to remain at an acceptable level, the reliability of each unit must be very close to one.

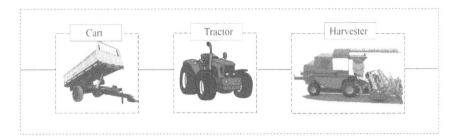

FIGURE 10.2 Three components in series situation.

There are three ways to improve a system:

1. Replace low-reliability components with more reliable components
2. Use of new equipment (with more reliability)
3. Use of redundant units

10.4.2.2 Parallel Systems

To introduce another type of component combination, we consider a system that can perform the necessary operations in the process in only one step and by one component of the system. Figure 10.3 shows an example of this system type.

Where each of the one to N machines alone will be able to execute the entire system mission. The system will stop only if all the machines fail.

So:

Probability of optimal operation = 1 – the probability of failure

Or by using the previously introduced symbols:

$$R_{system} = 1 - \text{Probability of total failure} \qquad (10.4)$$

R_i = Sound functioning probability of component i

$1 - R_i$ = Failure probability of component i

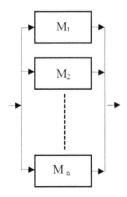

FIGURE 10.3 A parallel system.

According to the probability laws, the probability of failure of all components is:

$$R_{system} = (1 - R_1) \times (1 - R_2) \times \cdots \times (1 - R_n) \qquad (10.5)$$

Therefore, if the reliability of the components in a parallel composition is equal, its general formula is defined as follows:

$$R_{system} = [1 - (1 - R)^m]^n \qquad (10.6)$$

where m is the number of parallel components for each task, n is the number of tasks that each unit must perform, and R is the reliability of each component.

Example 10.2

In a parallel system, the same components are used independently, both machines and operators. Consider a hay bailer and a tractor to pull it. The farmer only needs a working tractor to pull the hay bailer. However, if he has two similar tractors so that both are available with the required 0.85 reliability at all times (Figure 10.4), the probability of having a tractor available to pull the hay bailer will be equal to:

$$R_{tractor} = [1 - (1 - 0.85)^2]^1 = 0.9775$$

In spite of the relatively low reliability of components like 0.6, a system with four parallel units can achieve a reliability of 0.9744. In some systems, components can include many different types, such as a human supports a machine or vice versa.

For the parallel configuration, the overall reliability of the system is greater than the reliability of each component.

Having a backup on a system can be implemented in several ways. Some common management options include:

1. Buy another machine
2. Borrow a backup if needed
3. Rent another unit
4. Get a reliable guarantee from the manufacturer

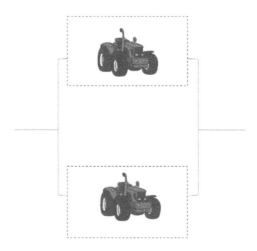

FIGURE 10.4   A parallel system for hay bailing.

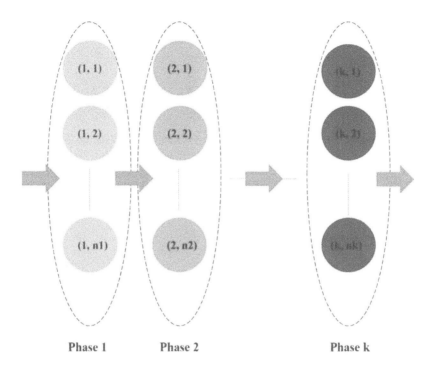

Phase 1   Phase 2   Phase k

FIGURE 10.5   Example of a series-parallel system.

10.4.3 SERIES-PARALLEL SYSTEMS

In a continuous system consisting of k phases, more than one machine (component) may be deployed to complete the mission in each phase; so, the system does not stop when a machine fails. In this case, it will be enough that at least one machine be healthy and usable in each phase for the system running [5]. The model for this combination is shown in Figure 10.5.

The number of each component is shown as (i, j) inside the circle. The system consists of k consecutive phases and a number of components are placed as parallel to each other in each phase. The symbol i represents the phase and the symbol j represents the component number inside phase i. There are n_i parallel components in each phase i.

So:

$$\text{The reliability of phase 1: } R_{s1} = 1 - \prod_{j=1}^{n1} q_{1,j} \qquad (10.7)$$

$$\text{The reliability of phase 2: } R_{s2} = 1 - \prod_{j=1}^{n2} q_{2,j} \qquad (10.8)$$

The reliability of phase k: $R_{sk} = 1 - \prod_{j=1}^{nk} q_{k,j}$ (10.9)

The reliability of system: $R_s = \prod_{i=1}^{k} R_{si} = \prod_{i=1}^{k}\left(1 - \prod_{j=1}^{nk} q_{i,j}\right)$ (10.10)

If the components inside each step are similar to each other, it means that we have:

$$q_{i.1} = q_{i.2} = \cdots = q_{i.n_1}$$

Then

$$\prod_{j=1}^{ni} q_{i,j} = q^{ni}$$ (10.11)

And as a result:

$$R_S = \prod_{i=1}^{k}(1 - q_i^{ni})$$ (10.12)

In a production line system consisting of series phases and parallel machines for each phase, we have:

Number of phase: 4
Number of similar machines in 1, 2, 3, and 4 phases = 3, 2, 2, and 3 machines, respectively,
Reliability of each machine in 1, 2, 3 phases = 0.8, 0.7, 0.75, and 0.9, respectively,
In these conditions, what will be the reliability of the whole system?

$$q_i = 1 - p_i$$

So,

$$q_1 = 1 - 0.8 = 0.2$$

$$q_2 = 1 - 0.7 = 0.3$$

$$q_3 = 1 - 0.75 = 0.25$$

$$q_4 = 1 - 0.9 = 0.1$$

As a result and based on the formula:

$$R_s = \prod_{i=1}^{k}(1 - q^{ni})$$

$$= (1 - 0.2^3) \times (1 - 0.3^2) \times (1 - 0.25^2) \times (1 - 0.1^3) = 0.845$$

10.4.4 GRAPHICAL TOOLS

10.4.4.1 Fault Tree Analysis

Fault tree (FT) is a top-down, structured methodology and deductive failure analysis in which the potential causes of a fault or undesired event are evaluated using Boolean logic to combine a series of lower-level events. The top event usually demonstrates a major incident, such as safety hazards or economic losses [6]. Although the top event occurs at the top of the tree, it is resolved downward in order to know further details about the system until the primary events causing the top event have been identified. In an FT, the binary nature of primary events is based on statistical independence. To represent relationships between events, AND-gates and OR-gates are most commonly used.

This is a top-down method in which one starts at a single point, branches downward, and then shows different states as logical symbols. An undesired event or fault is the starting point, and then the event can be resolved downward to identify the causes of the undesired event, which can then be examined to determine the causes of those events.

An FT can be evaluated quantitatively and qualitatively once it has been completed. In general, qualitative evaluation is derived by using Boolean algebra to determine the top event in terms of primary events. The occurrence probability of the top event is calculated for the quantitative part by considering the cut-set probabilities of the primary events.

A small FT can be manually evaluated; however, large and complex FTs are difficult to evaluate, and they should be analyzed by computer. FT analysis is frequently evaluated by several methods such as the analytical method, Monte Carlo simulation, and binary decision diagram. Because of the limitations in using Monte Carlo simulations to evaluate an FT, the analytical approach (e.g., minimal cut-sets determination) is commonly used.

10.4.4.2 Bayesian Networks (BNs)

In Bayesian networks (BNs), a set of variables are arranged to describe any probabilistic relationship between them [7–9]. Safety and reliability are two domains where BNs have been utilized increasingly. The BN is made up of two key parts: a directed graph as a qualitative element, and conditional probability tables as a quantitative element [10]. The BNs described in this study are based on discrete random variables. BNs are directed acyclic graphs (DAGs) where the edges must not make loops or cycles within the network. As mentioned above, in BNs, each variable is represented by the node of the DAG and the edges between the variables delineate direct informational or causal probabilistic relations between the linked variables. The directed edges are often presumed to signify causal relationships in the BN; however, probability theory cannot per se express causality, so edge directions do not necessarily indicate the causal effects.

When two nodes are connected by a directed edge, the "parent" is the node at the tail and the "child" is the node at the head, so the arrows points from cause or "the parent" node to effect or "the child" node based two common functions included intersection and union of variables, and establish a compact illustration of joint probability distributions.

If there are no parents for a node, a marginal probability table which represents the probability of each of the possible states will be defined for this node. A conditional probability table assigned to each node with parents specifies how each node states characterized by the conditional probability for the combination of every state of its parent(s). Then, the joint probability distribution of all the nodes would be calculated by these tables.

These networks are a pair of variables that are defined as $\langle\langle V,E\rangle,P\rangle$ where $\langle V,E\rangle$ represents the nodes and the edges of a DAG, respectively, and P gives the probability distribution over V which discrete random variables $V = \{X_1, X_2,\ldots, X_N\}$ are allocated to the nodes, while the edges E is the causal probabilistic relationship among the nodes.

Based on the conditional independence, as a result of the d-separation concept, and the chain rule, BNs indicate the joint probability distribution $P(U)$ of variables $U = \{X_1, X_2,\ldots, X_N\}$ of any BNs can be factorized as in Eq. (10.13)

$$P(X) = \prod_{i=1}^{n} P\big(X_i \mid pa(X_i)\big) \tag{10.13}$$

where $pa(X_i)$ is the parents of X_i in the BN, and $P(X)$ is the properties of the BN [11]. The main application of BNs in safety and reliability analysis is an inference engine that can update the prior probability of each event given new information or posterior probability, called evidence A. The new information gives operational data containing occurrence or nonoccurrence of the primary events as Eq. (10.14) [12].

$$P(X \mid A) = \frac{P(X,A)}{P(A)} = \frac{P(X,A)}{\sum_{X}(X,A)} \tag{10.14}$$

As illustrated in Figure 10.6, this BN includes a set of nodes $S = \{A, B, C, D, E, F\}$ in which interdependencies among the variables are signified by a set of edges. An outgoing edge from the prior node to the posterior node (such as $A \rightarrow B$) shows the interrelationship between these two nodes, so that the value of B is conditioned on the value of A. A is the parent node of B, C, and D, and B, C, and D are the child nodes of A. Generally, a BN structure is divided into three kinds: (1) root nodes are

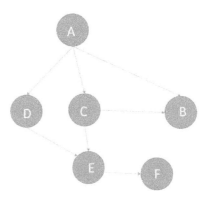

FIGURE 10.6   Directed acyclic graph of nodes A, B, C, D, E, and F.

defined as nodes without a parent, (2) leaf nodes are the nodes without a child node, and (3) intermediate nodes are the nodes with parent and child nodes. Since, in this BN, A has no parent node, it is considered the root node. B and F are the child nodes as they have not leaf nodes and C, D, and E are the intermediate nodes. The general expression of the full joint probability distribution can be calculated as Eq. (10.15).

$$P(A, B, C, D, E, F)$$
$$= P(F \mid E). P(E \mid C, D). P(D \mid A). P(C \mid A). P(B \mid C, A). P(A) \quad (10.15)$$

Either updating of probability or probability prediction can be used in this equation. In predictive analysis, the occurrence or nonoccurrence of a certain primary event can be determined by calculating conditional probabilities of the form $P(\text{accident}|\text{event})$ in which the occurrence probability of a particular accident is indicated. On the other hand, in updating analysis, the occurrence probability of a particular event can be identified by evaluating the form $P(\text{accident}|\text{event})$ in which the occurrence of a certain accident is shown [12, 13].

10.5 AVAILABLE APPROACHES FOR INCREASING RELIABILITY

To increase the reliability of safety-critical industrial systems, there are generally several basic methods that the following several methods are the most important.

- Utilizing redundancy technique:
  According to its engineering definition, by placing similar subsections in a system in parallel, we can guarantee the overall performance of the system in the event of an error.
- Finding the causes of a breakdown before they occur to eliminate these.
  Design the system in such a way that the methods of preventing possible failures are considered so that it is well possible to prevent breakdowns in the future.
- Maintenance optimization,
- Maintenance planning,
- By regularly monitoring the system status by software or user can automatically predict mandatory repairs and prevented possible breakdowns,
- Fault tree analysis,
- Sneak circuit analysis,
- Test design,
- Component and spare parts management, and
- Workload reductions are also effective ways to increase reliability.

10.6 CONCLUSION

Increasing reliability plays a major role in increasing the performance and efficiency of any industrial system. Given that repairing parts of a system is much cheaper than replacing that system, altogether predicting reliability will be very important in calculating its lifespan. This will justify the costs incurred in improving the reliability of a collection economically.

REFERENCES

1. Gunnarsson, C. and Hansson, P.A. 2004. Optimization of field machinery for an arable farm converting to organic farming, Agricultural Systems, Vol. 80, pp. 85–103.
2. Mobley, R.K. 2002. An introduction to predictive maintenance. Oxford, United Kingdom: Butterworth-Heinemann. 438 p.
3. Paman, U., Uchida, S. and Inaba, S. 2012. Operators' capability and facilities avail-ability for repair and maintenance of small tractors in Riau province, Indonesia: a case study, Journal of Agricultural Science, Vol. 4, No. 3, pp. 71–78.
4. Wireman, T. 2005. Developing performance indicators for managing maintenance. New York, NY: Industrial Press, Inc.
5. Shirmohammadi, A. 2002. Repair and maintenance planning. Iran: Arkane Danesh. 315 p.
6. Lewis, E.E. 1994. Introduction to reliability engineering. 2nd ed. New York, NY: John Wiley & Sons.
7. Insua, D.R., Ruggeri, F., Soyer, R. and Wilson, S. 2020. Advances in Bayesian decision making in reliability, European Journal of Operational Research, Vol. 282, No. 1, pp. 1–18.
8. Pearl, J. 1988. Probabilistic reasoning in intelligent systems: networks of plausible inference. San Francisco, CA: Morgan Kaufman.
9. Tong, X., Fang, W., Yuan, S., Ma, J. and Bai, Y. 2018. Application of Bayesian approach to the assessment of mine gas explosion, Journal of Loss Prevention in the Process Industries, Vol. 54, pp. 238–245.
10. Khakzad, N., Khan, F. and Amyotte, P. 2011. Safety analysis in process facilities: comparison of fault tree and Bayesian network approach, Reliability Engineering and System Safety, Vol. 96, No. 8, pp. 925–932.
11. Jensen, F.V. and Nielsen, T.D. 2007. Bayesian networks and decision graphs. 2nd ed. New York, NY: Springer.
12. Przytula, K.W. and Thompson, D. 2000. Construction of Bayesian networks for diag-nostics, Proceedings of IEEE Aerospace Conference, Vol. 5, pp. 193–200.
13. Afsharnia, F., Marzban, A., Asoodar, M. and Abdeshahi, A., 2020. Preventive main-tenance optimization of sugarcane harvester machine based on FT-Bayesian network reliability, International Journal of Quality & Reliability Management, Vol. 38, No. 3, pp. 722–750.

Index

181

For Product Safety Concerns and Information please contact our EU
representative GPSR@taylorandfrancis.com
Taylor & Francis Verlag GmbH, Kaufingerstraße 24, 80331 München, Germany

www.ingramcontent.com/pod-product-compliance
Ingram Content Group UK Ltd.
Pitfield, Milton Keynes, MK11 3LW, UK
UKHW021121180425
457613UK00005B/180